W9-CEO-503

THE SPACE TELESCOPE

THE SPACE TELESCOPE

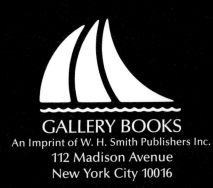

GALLERY BOOKS
An Imprint of W. H. Smith Publishers Inc.
112 Madison Avenue
New York City 10016

A Friedman Group Book

Published by
GALLERY BOOKS
An imprint of W. H. Smith Publishers, Inc.
112 Madison Avenue
New York, New York 10016

Copyright © 1987 by Michael Friedman Publishing Group, Inc.

All rights reserved. No part of this publication may be
reproduced, stored in a retrieval system, or transmitted, in
any form or by any means, electronic, photocopying, recording,
or otherwise, without the prior written permission of
the copyright owner.

ISBN 0-8317-7971-3

THE SPACE TELESCOPE
was prepared and produced by
Michael Friedman Publishing Group, Inc.
15 West 26th Street
New York, New York 10010

Editors: Sharon L. Squibb/Mary Forsell
Art Director: Mary Moriarty
Designer: Robert W. Kosturko
Layout: Fran Waldmann
Photo Editor: Philip Hawthorne
Production Manager: Karen L. Greenberg

Typeset by BPE Graphics, Inc.
Color separations by South Seas Graphic Arts Company Ltd.
Printed and bound in Hong Kong by Leefung Asco Printers Ltd.

DEDICATION

For Judith

CONTENTS

CONTENTS

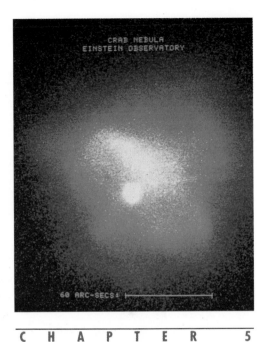

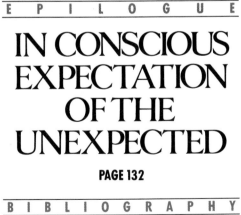
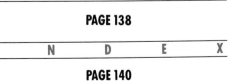

INTRODUCTION

In October 1608 a Dutch lens grinder named Hans Lippershey, a native of the city of Middelburg, applied to the government of the United Netherlands for a patent on a device to make distant objects appear closer. The government was interested in the military possibilities of this instrument and bought several but denied the inventor's request for a patent on the ground that "many other persons had knowledge of the invention." And, indeed, many people did. By April 1609, telescopes could be purchased in Paris, and by later that year they were available in the major cities of Germany, England, and Italy.

In July of that year, a forty-five-year-old professor of mathematics at the University of Padua—then under Venetian rule and well known throughout Europe as a center for free and erudite spirits— heard of the invention while visiting Venice. He returned to Padua and built a unique telescope that was capable of enlarging an object ninefold. The mathematician displayed his handiwork to an academic colleague with government connections. The grateful Venetians granted the professor, Galileo Galilei, a lifetime appointment at the university and an unprecedented raise.

Galileo continued to build more and better telescopes. By the end of the year he had made one capable of thirtyfold enlargement. Then, in January 1610, he turned his lenses to the night sky. To his own enormous surprise, he saw that the Moon is mountainous, the Milky Way consists of stars (many more stars than anyone had thought existed), Venus has phases like the Moon, and Jupiter has four lunar satellites of its own. In March he published his findings, initiating a revolution in astronomy.

"I'm perhaps something of an enthusiast," Princeton astronomer Lyman Spitzer says today, "but I think the space telescope will be the greatest improvement in optical astronomy since Galileo invented the telescope. It opens the way for seeing objects that we've never been able to see before— objects that have many important things to tell us about the nature, origin, and destiny of the universe." Spitzer is perhaps something of an enthusiast, but, if anyone deserves to be enthusiastic about the telescope, as one of its earliest and most tenacious supporters, he's the one.

While the idea of putting an optical telescope in space, where it would be undisturbed by Earth's obscuring, turbulent atmosphere, possessed an appealingly elegant simplicity, the project itself faced one obstacle after another. Support had to be marshaled within the astronomical community; funds had to be wrangled from Congress; and then, in this imperfect world, a near-perfect telescope had to be

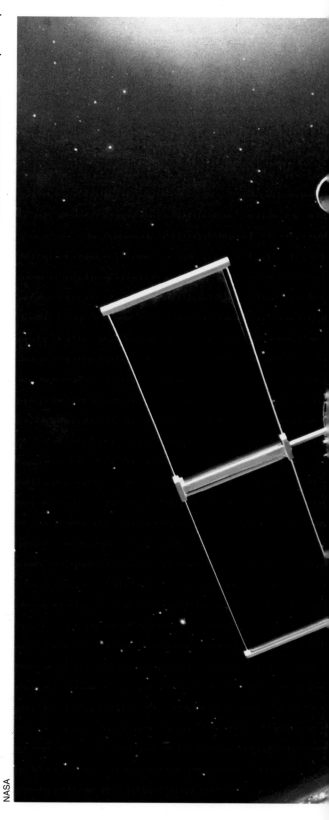

NASA

Offering scientists a window on the heavens of unprecedented power and clarity, the Hubble Space Telescope could lead to a revolution in astronomy comparable to the one Galileo began in 1610.

built, because distortions unnoticeable on Earth-based instruments would be all too apparent in space.

The whole process, from conception to launch, has taken more than forty years and cost a great deal of money with no payoff until after the launch. And what is the payoff? Fifteen years of detailed observations: of the planets that are our fellow travelers about the Sun and, perhaps, of a yet unknown planet journeying around some other sun; of the birth, life, and death of stars; of galaxies so distant they hold the key to the overall structure and origin of the universe; and of other objects still unknown and unimaginable to us.

BEFORE THE DRAWING BOARD

NASA

Earth's atmosphere is no ally of the observational astronomer. At the same time that it permits life to flourish on this planet, the air that surrounds us seals out a great deal of the energy coursing through space. The atmosphere also greatly limits the clarity of the energy that reaches us on Earth. Consider the effect of the atmosphere on visible light. When children sing about the twinkling of the stars, they are not celebrating some inherent quality of these celestial objects, but rather the effect of atmospheric turbulence. And while the glittering, diamondlike result may hardly seem like a disfigurement to the casual star-gazer, it is a serious obstacle to the scientist.

Astronomers measure the apparent size of celestial objects in terms of degrees of angle, which are further divided into arc minutes, 60 of which make up 1 degree $(60' = 1°)$, and arc seconds, 60 of which make up 1 arc minute $(60'' = 1')$. The entire sky, from horizon to horizon, consists of 180°, with the largest object in the night sky, the Moon, only half a degree, or 30 arc minutes across $(.5° = 30')$. Unaided, the human eye cannot discern details any finer than one or two arc minutes. This limits our ability to observe clearly even the largest planet in our Solar System, Jupiter, which at its closest approach to Earth appears just under an arc minute. Resolution improves considerably with any sort of telescope, and the best ground-based telescopes can resolve objects as small as 1 arc second in size.

On the cosmic scale astronomers work on, however, 1 arc second can be a severe limitation. It means they cannot distinguish individual stars in a cluster, clearly see a bright object's faint companion,

In their search for locations where atmospheric distortion is least likely to occur, astronomers in the past few decades have built observatories in the Chilean Andes, the Canary Islands, and Hawaii—like this infrared telescope atop Mauna Kea volcano (previous page).

The 200-inch (5-meter) Hale telescope at Mount Palomar, California, was the largest ever when it was completed in 1948 (right). In the race for telescopes with high resolution, this one was surpassed in the late 1970s by a Soviet-built 236-inch (5.9-meter) telescope in the Caucasus.

Courtesy The Carnegie Institution of Washington

The William Herschel Society/Michael Tabb

William Herschel (left), who gained worldwide fame for his sighting of the planet Uranus in 1781, discovered the existence of infrared radiation in 1800.

measure the subtle movement of a body that might indicate the presence of a planetary system, or make any of a large assortment of key observations.

Modern astronomers have expended great effort to overcome this 1-arc-second barrier. They have searched Earth for elevations where the air is less subject to disturbance as well as less contaminated by industrial pollution and the obscuring glare of cities. As a result, in the past two decades, new large telescopes have been placed on mountaintops in Chile, Hawaii, and the Canary Islands. The placement of these telescopes has given astronomers better resolutions as well as created dramatically more difficult working conditions. However, while these telescopes have lessened the obstacles the atmosphere poses, they have not eliminated them.

And, at the same time that the atmosphere distorts visible light, it almost entirely screens out many of the other forms of energy traveling through the universe. These forms of energy, contemporary astronomers believe, contain important clues about the birth, structure, and fate of the cosmos. William Herschel, the British astronomer best remembered for finding the planet Uranus, discovered the first of these invisible energies in 1800. It had long been known that when sunlight passes through a glass prism it is divided into the colors of the spectrum—red, orange, yellow, green, blue, indigo, and violet. Herschel, attempting to measure the temperatures of the different bands of color, noticed that a thermometer placed next to the red band became hot even though no visible light was striking it. He called the new form of energy *infrared radiation.*

By the end of the nineteenth century, scientists had determined that visible light and infrared radiation are part of a large group of related forms of energy, known as the electromagnetic spectrum, all traveling at the same speed and differing only in wavelength and frequency. Visible light occupies the central position on this spectrum. Short wavelength forms are, in order of decreasing wavelength, ultraviolet light, X rays, and gamma rays. Long wavelength forms are, in order of increasing wavelength, infrared radiation and radio waves.

The investigation of most of these forms of energy as astronomical phenomena could not take place until the space age. While most radio waves and some infrared light reach the surface of our planet, gamma rays, X rays, and ultraviolet light are virtually completely excluded. The little ultraviolet light that does reach our planet's surface is responsible for the tanning and burning effect of the Sun on human skin.

A few pre-space scientists had given some thought to how the screening and distorting effects of the atmosphere might be avoided.

NASA

Werner Von Braun (above), who designed the *V-2* rocket with which Nazi Germany bombarded England during World War II, also developed the *Redstone* rocket, which lofted the first American astronaut, Alan Shepard, into space in 1961.

Modern astronomy has developed a variety of instruments that enable scientists to observe the different forms of energy in the electromagnetic spectrum. Note: this chart (right) is somewhat dated and refers to the space telescope as the LST (Large Space Telescope), a name that was changed in the mid-1970s.

NASA

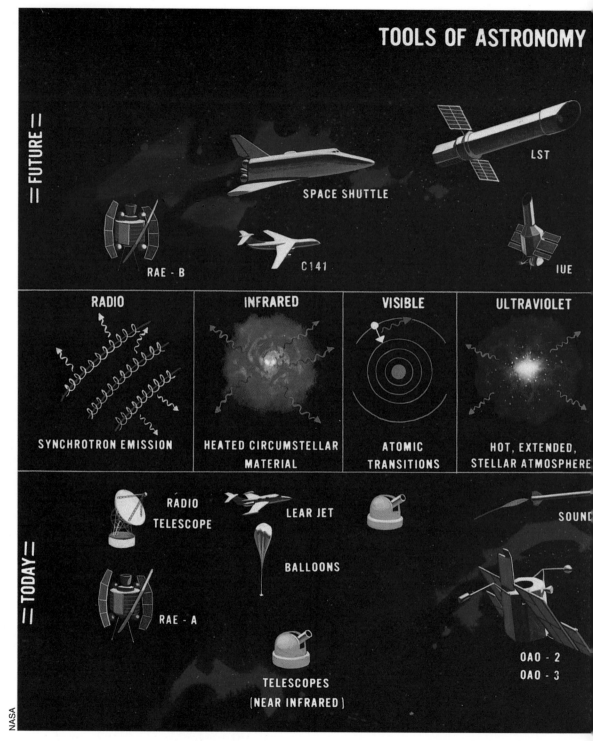

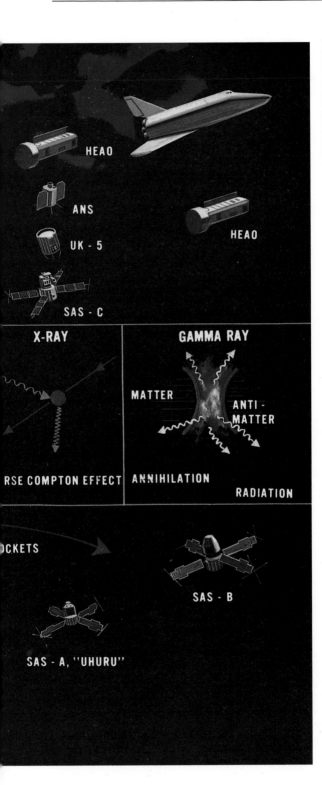

HEAO

ANS

UK - 5

HEAO

SAS - C

X-RAY

GAMMA RAY

MATTER

ANTI - MATTER

RSE COMPTON EFFECT

ANNIHILATION

RADIATION

OCKETS

SAS - B

SAS - A, "UHURU"

Back in the 1920s, German rocketry pioneer Hermann Oberth suggested that a telescope in space would enjoy certain inherent advantages over its earthbound brethren. Oberth's ideas were, in general, highly visionary and extremely impractical for his time, but not without their effect. His books inspired at least one German adolescent, previously an indifferent student, to learn mathematics in order that he might realize some of Oberth's grand schemes. The boy's name was Werner von Braun. And the V-2 rockets built by von Braun (with Oberth as one of his advisers) during World War II to assist in the Nazi bombardment of England were instrumental in turning the thoughts of the United States to the other possible uses of outer space.

As the war was drawing to a close, the United States Air Force, intrigued by the V-2s, commissioned a study on the uses of missiles. Among its conclusions, the study noted that satellites in orbit around Earth would be of tremendous military and scientific value. The Air Force commissioned a second study called Project RAND (an acronym for "research and development"). Among those asked to participate was a young astrophysicist named Lyman Spitzer. Although he already had a more than full-time position with a civilian research group investigating submarine warfare, Spitzer readily accepted. For many years, he had been more interested in what lay beyond Earth's atmosphere than in how to navigate, or keep others from navigating, beneath its oceans.

In 1930, while still a prep-school student, Spitzer attended a lecture by Henry Norris Russell, the head of the Princeton Observatory, which would determine the ultimate if not the immediate direction of his future career. Although Spitzer was inspired by Russell's lecture to read widely in the field of astronomy, particularly the writings of the English astrophysicist Sir Arthur Eddington, when he attended Yale he majored in physics and completely avoided all courses in star science. As a graduate student, he set off to England to study nuclear physics but ended up working with Eddington at Cambridge University. When he returned to the United States, Spitzer completed his doctoral in physics at Princeton under the tutelage of Russell. The physics department, however, hesitated to accept a thesis characterized by the presence of so much astronomy. After asking Russell's advice, the department decided to award Spitzer Princeton's first advanced degree in astrophysics in 1938.

In his paper written for the Air Force's Project RAND in 1946, Spitzer discussed several potential astronomical uses for satellites. He suggested sending two small telescopes sensitive in the ultraviolet range of the spectrum into orbit. He also used the paper to recommend that a large optical telescope should be sent up in the satellite. He would spend much of the next forty years aggressively promoting this project both in the groves of academe and the halls of Congress. "It was quite obvious to me what the tremendous advantages of a

Under the direction of Martin Schwarzchild, Stratoscope II (below) extended to a height of 80,000 feet (24,400 meters) to observe Jupiter, Uranus, and various star systems.

NASA

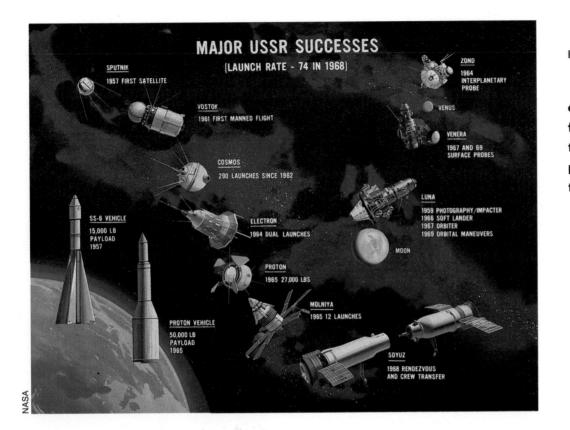

MAJOR USSR SUCCESSES
(LAUNCH RATE - 74 IN 1968)

SPUTNIK
1957 FIRST SATELLITE

ZOND
1964 INTERPLANETARY PROBE

VOSTOK
1961 FIRST MANNED FLIGHT

VENUS

VENERA
1967 AND 69 SURFACE PROBES

COSMOS
290 LAUNCHES SINCE 1962

LUNA
1959 PHOTOGRAPHY/IMPACTER
1966 SOFT LANDER
1967 ORBITER
1969 ORBITAL MANEUVERS

SS-6 VEHICLE
15,000 LB PAYLOAD
1957

ELECTRON
1964 DUAL LAUNCHES

MOON

PROTON
1965 27,000 LBS

PROTON VEHICLE
50,000 LB PAYLOAD
1965

MOLNIYA
1965 12 LAUNCHES

SOYUZ
1968 RENDEZVOUS AND CREW TRANSFER

NASA

The Soviet launch of *Sputnik* in 1957 led the United States to accelerate its efforts to launch an artificial satellite. By the late 1960s, the Russians had compiled an impressive record in space as noted in this chart (left).

space telescope might be, and I worked a bit on physically what one could do," he says, recalling the 1946 paper. And, gradually, the United States did take a few small, hesitant steps into space. In 1946, a V-2 rocket that the Americans had captured from the Germans at the end of World War II made the first ultraviolet observations of the Sun. But, while the German rockets were used for a variety of astronomical tasks, postwar America chose not to embark on a large-scale effort to explore space.

In 1947, Spitzer returned to Princeton to assume Russell's former position as the head of the observatory. Did he believe that the space telescope was an idea whose time would ever come? "I was con-

vinced this was the way for astronomy to go and that someday it would go that way," he recalls. "As to the time scale, I had no very definite ideas in the beginning. While rocket research had a big boost during the Second World War, it wasn't clear how fast it would go after that."

In the absence of a national space program, Spitzer and his Princeton colleagues concentrated on less ambitious projects. "The program of our observatory was designed to push the general goal of a space telescope by getting involved with space research on a much smaller scale," he says. The specific program Spitzer aided was Martin Schwarzschild's effort to send telescopes up in a balloon. In 1957 and 1959, Schwarzschild conducted a se-

ries of observations using funds provided by the United States Navy and the National Science Foundation. An automated 10-inch (25-centimeter) telescope named Stratoscope I obtained extremely sharp photographs of the Sun at approximately 20 miles (32 kilometers) above the densest part of the Earth's atmosphere. In the next decade, Stratoscope II took photographs of planets and star systems with a 36-inch (91-centimeter) telescope and a resolution approaching 0.1 of an arc second.

Spitzer's original plan had been to exhaust the possibilities of balloons in astronomical research before attempting to raise the money needed to put a satellite in orbit. But the launch of *Sputnik 1* by the

Soviet Union in October 1957 seriously altered that schedule.

"Someone from the Air Force phoned us and said we really ought to support their satellite project," Spitzer remembers. "While our schedule didn't really call for us to get involved with satellites until the Stratoscope was over, we accelerated our plans and did both at the same time." Goaded by the Soviet achievement, the United States formed its own National Aeronautics and Space Agency (later rechristened the National Aeronautics and Space Administration, or NASA) in July 1958.

The first NASA satellites intended for astronomical observation, the four Orbiting Astronomical Observatories, had mixed records. They had been built to survey the ultraviolet range, the domain of very hot and very young stars, but the first one did not work because an electrical battery malfunctioned, and the third was

dumped into the Indian Ocean after experiencing rocket failure. The second and fourth, however, launched in 1968 and 1972, were spectacular successes. The final of these satellites, named *Copernicus* a year after launch to commemorate the five hundredth anniversary of the Polish astronomer's birth, remained in orbit for nine years. It carried a 36-inch (91-centimeter) telescope with instrumentation that enabled it to measure ultraviolet spectral lines. The device had been designed by a group from Princeton led by Spitzer. "*Copernicus* opened up completely new windows on the gas between the stars, which plays an important role in the overall life cycle of a galaxy," observes Spitzer. "The way the gas behaves in interstellar space and gets involved in the formation of new stars is a fascinating question."

The *International Ultraviolet Explorer (IUE),* launched in 1978, is a

NASA's third Orbiting Astronomical Observatory (far right) never made it into orbit due to rocket failure. However, the fourth observatory, *Copernicus* (right), was a spectacular success.

NASA

NASA

similar success. It is a joint project of the British Science Research Council—which built and designed the camera—the European Space Agency (ESA)—which contributed the solar panels—and NASA—which did everything else. *Copernicus* and *IUE* have radically changed astronomical ideas about interstellar gas. Rather than being evenly distributed throughout space as had been previously believed, the gas now appears to exist in dense clouds permeated by holes blasted out by exploding stars. In the words of one astronomer, "The structure of the invisible gas in our galaxy resembles a swiss cheese."

While *Copernicus* and the *IUE* were being designed and built, the idea of launching a large optical telescope into Earth orbit was also gaining momentum. In 1962, a group of scientists assembled for NASA by the National Academy of Sciences issued a report recommending the development of a large space telescope as a logical long-range goal of the American space-science program. The recommendation was repeated by a similar group three years later, which concluded that "a space telescope of very large diameter... requiring the capability of a man in space, is becoming technically feasible, and will be uniquely important to the solution of the central astronomical problems of our era."

Soon after that, the National Academy formed a committee chaired by Spitzer to define the scientific objectives of such a telescope. The committee report, published in 1969, called the proposed device the Large Space Telescope and determined that it would have a primary reflecting mirror with a diameter of 120 inches (3 meters). According to the report, the space telescope could be "visualized as a permanent space observatory" that "would have to be manned at relatively frequent intervals." Scientifically, it "would make a dominant contribution to our knowledge of cosmology—to our understanding of the content, structure, scale, and evolution of the universe." The report also noted that such a space telescope would also provide important and decisive information for many other fields of astronomy.

The report won NASA approval, and in 1971 the space agency launched feasibility studies on the project. The next year, Charles R. O'Dell, director of the University of Chicago's Yerkes Observatory, was appointed project scientist. The Marshall Space Flight Center in Huntsville, Alabama, was designated the lead center for the mission from inception through launch. Additionally, the Goddard Space Flight Center in Greenbelt, Maryland, was put in charge of building the five scientific instruments that would accompany the telescope into space as well as directing postlaunch operations. The only remaining obstacle before design and construction could begin was the need to obtain funding from Congress.

Spitzer's committee had already begun to lobby among its fellow scientists. "It took time for the astronomy community to be convinced of the promise, the potential, and the value of the space telescope," Spitzer admits. "I think the key func-

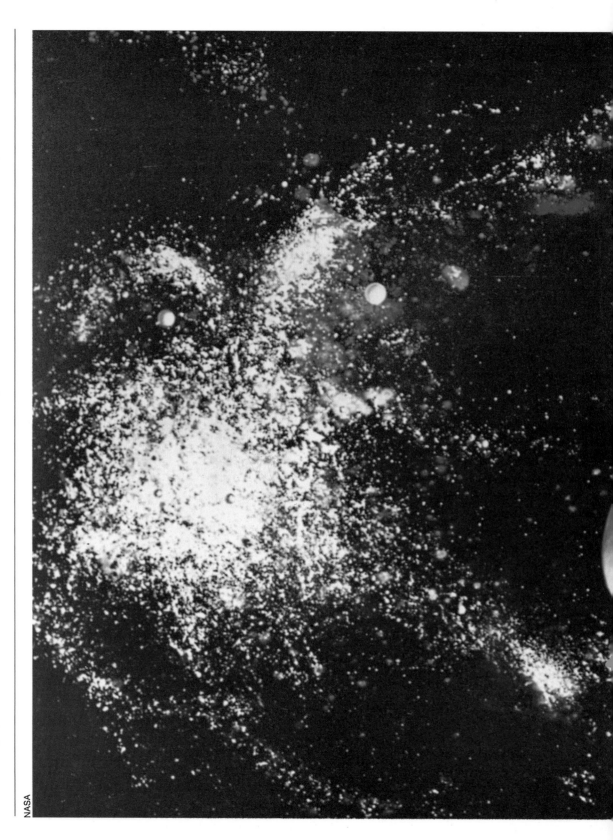

NASA

In 1971, NASA began feasibility studies on the space telescope—then officially known as the Large Space Telescope (shown here in a 1974 artist's conception). The next year NASA appointed a project scientist and designated the Marshall Space Flight Center in Huntsville, Alabama and the Goddard Space Flight Center in Greenbelt, Maryland as lead research centers. Congress, however, did not approve funding for the project until 1977.

tion that our committee had during the several years of its existence was to provide communications with research astronomers in a variety of disciplines. We organized a series of small conferences to discuss what could be done with the space telescope in cosmology and other fields. The people who attended came away, I think, with a very real conviction that the space telescope would work, that it really would be a revolutionary tool."

Whether the response really was as enthusiastic as Spitzer recalls seems questionable. In 1971, a committee of twenty-three astronomers assembled by the National Academy and chaired by Jesse Greenstein of the California Institute of Technology had listed the space telescope as the ninth on a list of recommended projects, of which only the first four were given top priority. Spitzer and his colleagues appear to have been guilty of showing too much confidence at that stage, an attitude that served them particularly poorly when dealing with Congress. When the idea for the space telescope was first presented to the House subcommittee that oversees the NASA budget, discussion lasted five minutes and centered almost exclusively on the question of how much the project would cost. Citing the lukewarm account of the project in the Greenstein report, the House ultimately refused all funding.

The supporters of the space telescope realized that some serious lobbying was in order if their project was ever to be realized. Before going back to Congress, they attempted to close ranks within the

NASA

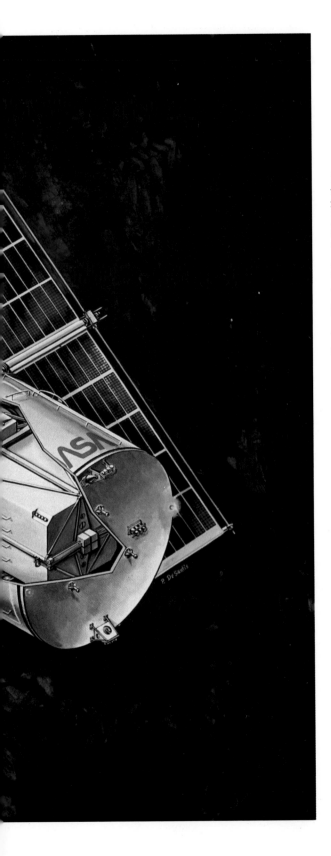

Courtesy Princeton University/Robert Matthews

In a paper he wrote in 1946, Lyman Spitzer (above) first noted the potential of a telescope in orbit around the Earth. Throughout the 1960s and 1970s, he was an enthusiastic lobbyist for the telescope (left), both with Congress and the scientific community.

scientific community, specifically within the Greenstein committee. All twenty-three members of the committee signed a brief statement in support of the project: "In our view, Large Space Telescope has the leading priority among future new space astronomy instruments." Then the telescope's supporters returned to Washington.

"Both NASA and Congress were really astounded by the breadth and intensity of the support that became apparent," recalls Spitzer. "Projects that come up for government support are usually only endorsed by the few people who will make use of that particular apparatus. In this case, there was scarcely an astronomer who wasn't visualizing himself as using the space telescope." The effort finally paid off in 1977, when Congress authorized construction of the project.

Between the time of the studies at NASA and the legislative battles on Capitol Hill, the space telescope had undergone a few alterations. Its name was no longer Large Space Telescope, but Hubble Space Telescope, named for one of the leading American astronomers of this century. Its size had been slightly scaled down: The diameter of the reflecting mirror had been reduced from 120 inches (3 meters) to 94 inches (2.4 meters). Its life expectancy was determined to be fifteen years, although with replaceable parts it might yet prove to be the permanent observatory envisioned in the late sixties. This goal was even closer, for it would be placed into low Earth orbit with periodic repair visits made by astronauts on the space shuttle.

TO BUILD A TELESCOPE

"The object of any large astronomical observatory is to take high-resolution pictures of very faint objects," explains Terrence Facey. Facey is a physicist with Perkin-Elmer, the Danbury, Connecticut, firm that in October 1977 was awarded the contract to build the optical system and the fine-guidance sensors for the space telescope by NASA's Marshall Space Flight Center.

Facey, an Englishman, has been building telescopes to be launched into space since he joined Perkin-Elmer after graduating from the University of London in 1967. He worked briefly on *Copernicus* and then joined the team that built the solar observatory for *Skylab*. Since 1972, he has been involved with the space telescope, first with the preliminary planning, then with the actual construction. Although most contemporary astronomers

are, like Spitzer, physicists who became fascinated with the physics of the heavens, and in spite of his long experience building astronomical instruments, Facey is not deeply interested in astronomy. "I have what I would call a popular science interest in astronomy," he says, "but I was never tempted to enter the field. I think I like hardware a little too much."

Building the hardware for the space telescope presented a number of unusual problems, not so much in the basic design of the telescope as in meeting the specifications the telescope had to fulfill. When most people think of a telescope, they imagine a refracting telescope, which has a lens at the top that focuses light for the viewer, who gazes at the image formed at the lower end. This is the sort of telescope Galileo used, and it is still used today by scientists, notably at

Courtesy Perkin-Elmer

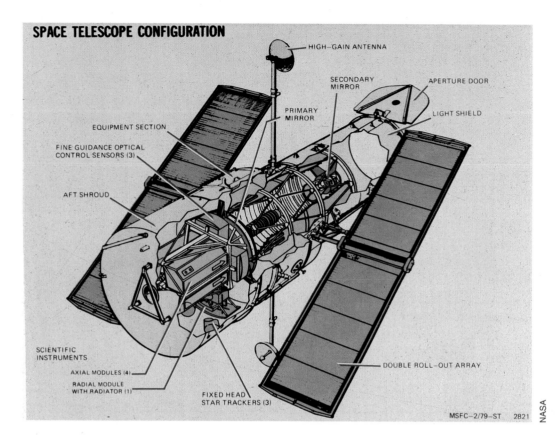

SPACE TELESCOPE CONFIGURATION

HIGH–GAIN ANTENNA

SECONDARY MIRROR

APERTURE DOOR

PRIMARY MIRROR

LIGHT SHIELD

EQUIPMENT SECTION

FINE GUIDANCE OPTICAL CONTROL SENSORS (3)

AFT SHROUD

SCIENTIFIC INSTRUMENTS

AXIAL MODULES (4)

RADIAL MODULE WITH RADIATOR (1)

FIXED HEAD STAR TRACKERS (3)

DOUBLE ROLL–OUT ARRAY

MSFC–2/79–ST   2821

NASA

NASA

An artist's rendering (previous page) provides us with a glimpse of the space telescope's many possible uses. Via the Space Shuttle, astronauts will be able to service and repair the telescope's giant curved mirrors and cameras.

The space telescope's optical system—primary and secondary mirrors and graphite epoxy metering truss—and fine guidance sensors were manufactured by the Danbury, Connecticut, firm of Perkin-Elmer (above). The space telescope's mirrors (right) are made of Corning's Ultra Low Expansion glass, which does not expand or contract with changes in temperature.

the Yerkes Observatory, which has a refractor with a 40-inch (1-meter) lens weighing almost .25 of a ton (0.2 metric tons). The Yerkes refractor, the largest of its type in the world when it was completed in 1897, is likely to remain the largest. Because a refractor's lens can only be supported on its edge, a bigger lens would sag, and it would then be unusable.

The space telescope, however, like virtually all of the large telescopes built in this century, is a reflecting telescope, a type first employed by Sir Isaac Newton in 1672. It consists principally of a curved primary mirror with a 94-inch (2.4-meter) diameter situated at the end of the telescope tube and a 12.5-inch (32-centimeter) secondary mirror, which then redirects the light back to a small hole in the primary mirror. The image comes to a focus several feet behind the primary mirror, where small mirrors direct it to the fine-guidance sensors, which point the telescope to whichever of the five scientific instruments (two cameras, two spectrographs, and a light-metering device) the observing astronomer is using at the time. Technically, the space telescope is described as a Ritchey-Chretien type of Cassegrain optical system.

Several problems are associated with such demanding requirements. Facey explains, "When you combine those two needs—the need to see with very high resolution and the need to see very faint objects—you realize that you need a high-quality optical system to get high resolution, and you need a very stable optical system because

Pictured (at left) is the initial grinding of the primary mirror. To prevent the mirror from sagging in space, engineers devised a way of simulating the lack of gravity here on Earth.

you'll need a significant length of time to build up enough light from this very faint object." The space telescope resolves the problems presented by using the best optical mirror ever built—a support structure for the telescope that will neither expand nor contract in the extremes of space temperature—and fine-guidance sensors that will point and lock the telescope to stars so faint that no astronomer has yet seen them.

There are reflecting telescopes whose primary mirrors have larger diameters than the 94-inch (2.4 meter) one of the space telescope. The 200-inch (5-meter) reflecting telescope that sits atop Mount Palomar in California was the largest from its completion in 1948 until the late 1970s, when the Soviet Union placed a 236-inch (5.9-meter) tele-

scope almost a mile and a half above sea level in the Caucasus. (The Soviet effort, however, has been plagued by structural problems that have limited the quality of data it has produced.) But these instruments, for all their size, are limited by Earth's atmosphere and so do not require as finely shaped or polished a mirror as the space telescope would.

Perkin-Elmer made the space telescope's mirror out of Ultra Low Expansion glass manufactured by the Corning Glass Company. As its name implies, this type of glass does not expand or distort as a result of temperature changes. Another advantage of the glass is its weldability, which means thin slabs can be fused together to form a disk with square-celled, monolithic, honeycomb cores. Such a disk weighs one-fourth as much as one of solid glass, and the space telescope's nearly 8-foot (2.4 meter) primary mirror weighs only 1 ton (0.9 metric tons). Mount Palomar's solid-glass 200-inch (5-meter) mirror weighs 14 tons (12.7 metric tons), considerably more than the 11.25 tons (10.2 metric tons) of the entire space telescope—scientific instruments, solar panels, and all.

"Only when you want to put a telescope in space is there merit in spending the additional resources necessary to produce a nearly perfect mirror," says Facey. "It was necessary that the mirror be smooth because any lack of smoothness, the very short ripple that might occur and routinely does occur on the surface of a mirror when you hand polish it, can destroy performance in the ultraviolet. Since there had never previously been a need to polish a mirror of this size to that kind of tolerance, the technology to do it, obviously, was not in place." Nor was the technology in place to build so finely shaped a mirror. Perkin-Elmer had built 1-foot (0.9-meter) spheres to similarly stringent requirements but never an 8-foot (2.4-meter) parabola.

There was, in addition, another problem with regard to shaping the mirror, and that was gravity: It is present here on Earth, where the mirror was made, but is absent in space, where the telescope will function. If nothing were done to account for this discrepancy, the mirror would sag when it left Earth's gravitational field, just as it was supposed to start working. The solution the Connecticut firm devised was to simulate zero grav-

NASA

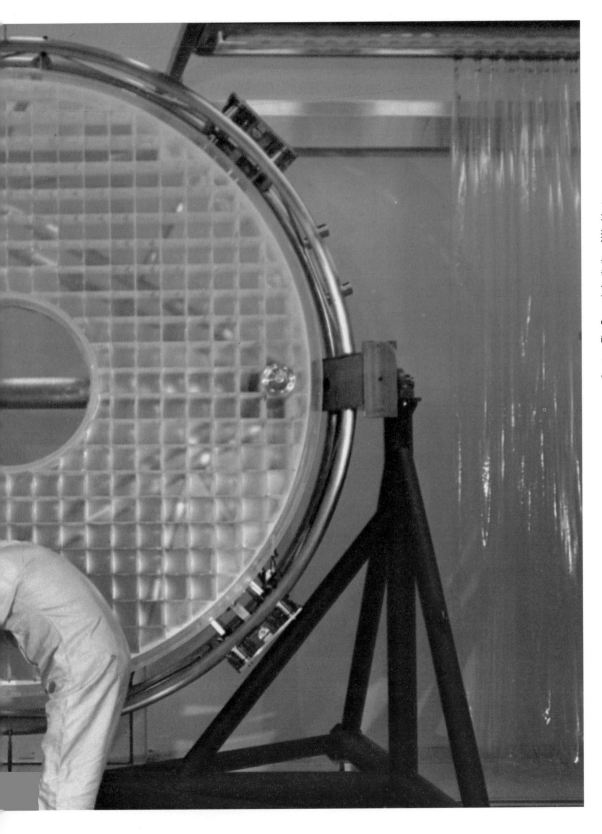

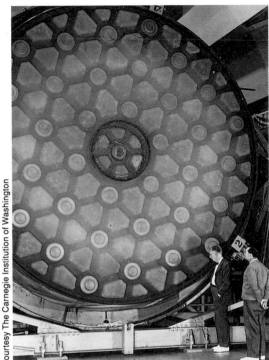

Courtesy The Carnegie Institution of Washington

The space telescope's 94-inch (2.4-meter) primary mirror (left), a disk of thin glass slabs fused together to form square-celled honeycomb forms, weighs only 1 ton (0.9 metric tons). In contrast, the 200-inch (5.1-meter) solid glass mirror on Mount Palomar (above) weighs 14 tons (12.7 metric tons).

The space telescope's primary mirror (right) is the finest large mirror in the world. "If the Pacific Ocean were as smooth," remarks one astronomer, "the highest wave would only be 0.001 inches [.025 millimeters] high."

Courtesy Perkin-Elmer

ity on Earth. Fifty engineers worked for three years to create 138 rods of support for the mirror, with each rod exerting a different, computer-determined force. In doing so, they succeeded in creating an environment that would differ from the orbital environment by less than 0.0033 wave of distortion.

Computers were also key in the polishing of the mirror. Not to imitate the effects of space but to avoid the inevitable distortions that would have been caused by polishing the surface by hand. "Those two requirements, good smoothness and a very accurate figure, pushed us to develop a computer-controlled polishing technique whereby the mirror is entirely polished by what is basically a numerically controlled machine," says Facey. When the polishing was complete in May

1981, Perkin-Elmer had made the finest large mirror in the world. "If the Pacific Ocean were as smooth as the space telescope mirror," remarks one astronomer, "the largest wave would only be a 0.001 inch [0.00254 centimeters] high."

Meanwhile, since November 1977, another team of engineers at Perkin-Elmer had been devising a method of coating the polished surface. A machine with rotating arms was created to clean the mirror, a process that took five days and consumed 2,400 gallons (9,084 liters) of hot, deionized water. Then, on 5 December 1981 the mirror was placed in a vacuum chamber, slowly heated to 250°F (121°C), spun at a high velocity, and coated with a 2.5-microinch (.0635 microns) layer (a microinch is 0.000001 of an inch) of aluminum,

Specially outfitted in masks and suits, Perkin-Elmer technicians (right) inspect the reflective surface of the newly coated primary mirror. Coating and cleaning the mirror was a long and delicate procedure, but the rewards of a clean, well-engineered mirror were worth the effort. The space telescope will enable astronomers and scientists to see seven times further into space than with conventional telescopes.

NASA

A computerized color-coded topographic map (right) reveals the mirror's irregularities before it was polished. White represents the optimum surface evenness, while blue and red correspond to its highs and lows.

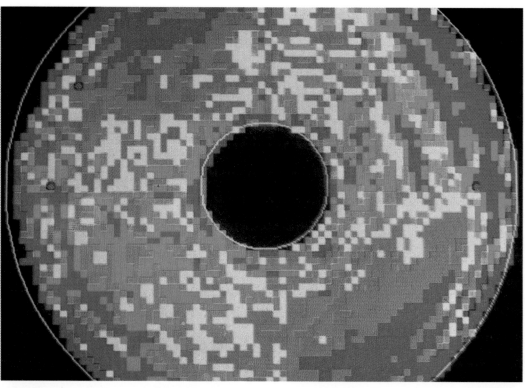

NASA

A computerized color-coded topographic map (right) illustrates the mirror's near-perfect surface after it was polished.

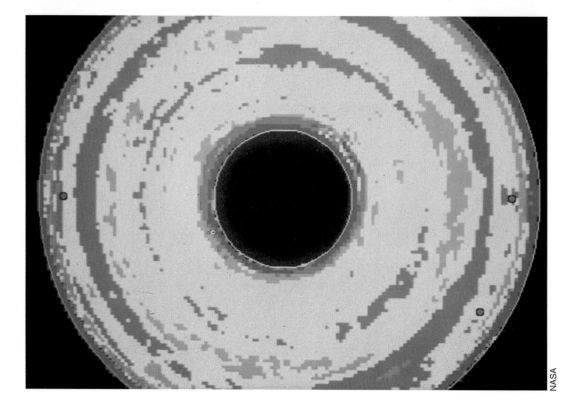

NASA

A technician scrutinizes the space telescope's primary mirror (right), which was coated with aluminum and magnesium fluoride in 1981.

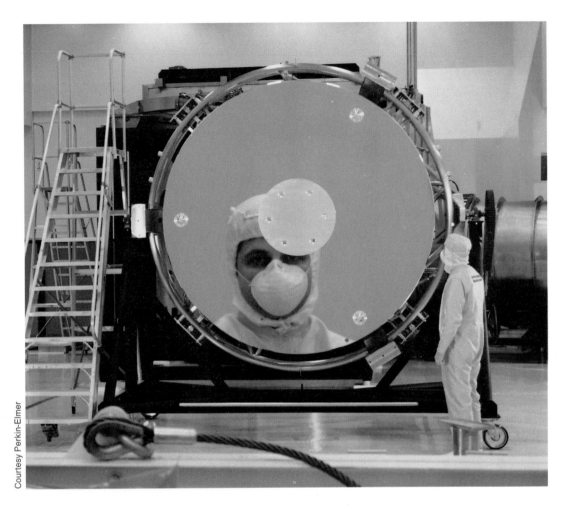

Courtesy Perkin-Elmer

which was then immediately protected from oxidation with a 1 microinch (0.000025-centimeter) layer of magnesium fluoride. In four minutes, a task planned for four years was over.

The truss built by the Boeing Aircraft Company to hold the space telescope's pair of exquisite mirrors in place had to share at least two of the qualities of those reflecting surfaces. It had to be lightweight and also be resistant to expansion and contraction at varying temperatures. This was accomplished by using graphite-reinforced epoxy, a new material that NASA had

started to investigate in the early 1970s. The result is a truss 210 inches (5.3 meters) long and 115 inches (2.9 meters) in diameter that weighs only 252 pounds (114 kilograms). The truss at Mount Palomar, by contrast, weights 120 tons (108 metric tons). And, while the space telescope is expected to experience variations in temperature of as much as 280°F (137°C) in space, the truss will not alter its size by more than 118 microinches (3 microns).

Following its relatively unprecedented use in the space telescope, graphite epoxy has become an in-

creasingly popular material for a variety of consumer uses—particularly in sporting goods. Graphite-epoxy golf clubs, tennis rackets, and bicycles have recently hit the market, although Facey says that there is no relationship between the material's terrestrial and space uses. "The application of graphite epoxy to golf clubs has nothing to do with the thermal ability for which the material was exploited on the space telescope," the physicist asserts. "Its light weight and high stiffness may lend something to golf clubs, but, not being a golfing enthusiast, I'm not sure what."

The final crucial responsibility of Perkin-Elmer was building the three fine-guidance sensors, two of which will be used to locate and lock onto targets, while the third functions as a sixth scientific instrument (with a scientific instrument team of its own) measuring the relative positions of celestial objects to within 0.002 arc seconds, about five times more accurately than typical ground based observations. Each fine-guidance sensor has two encoders and four photomultipliers that count light particles, or photons, received by two pairs of Koesters prisms. (A Koesters prism consists of two right triangles joined along a common leg to an equilateral triangle.) Light from a target star produces an interference pattern in each prism, and, should the space telescope move off target, both patterns change, one brightening, the other darkening. By monitoring these patterns, the position of the telescope can be kept exactly on target every second.

These sensors will enable long exposures of faint objects by keeping the space telescope aligned within 0.007 arc seconds for up to twenty-four hours. It is an awesome possibility, one that has inspired scientists and science writers to create an oddly related series of metaphors. The sensors possess "a precision that would let a laser in Washington pierce a dime in Boston," reported *Science* magazine. In the words of *Physics Today,* the sensors could " 'see' a row of dimes on Michigan Avenue in Chicago from the White House." Harvard-based astronomer and physicist Alan

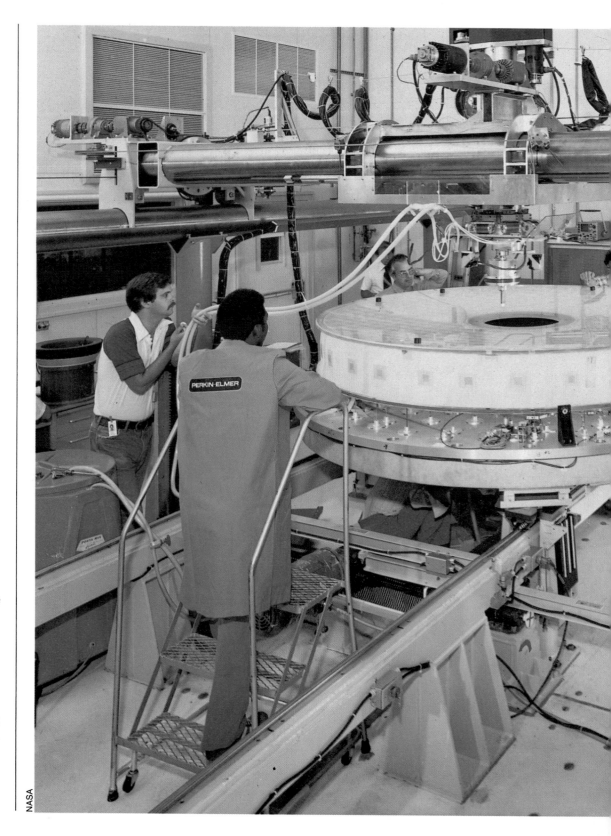

NASA

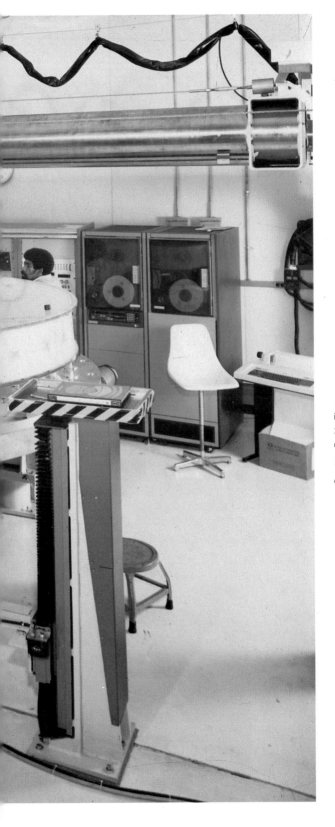

Courtesy Perkin-Elmer

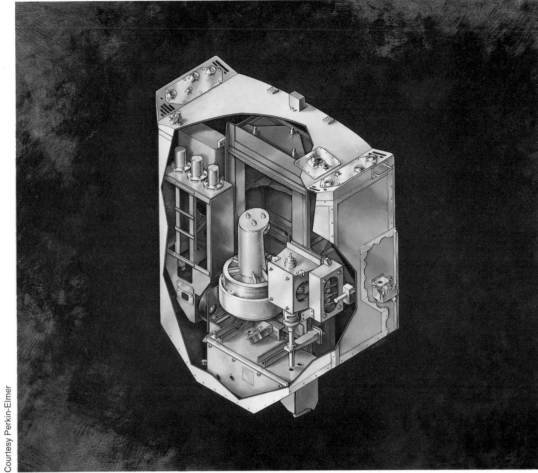

A cross section of a fine guidance sensor (above) shows an instrument "with a precision that would let a laser in Washington pierce a dime in Boston," as reported by *Science magazine*.

P erkin-Elmer developed a computer system (left) to polish the mirror in order to avoid the distortions inevitable in hand polishing.

Lightman describes the sensors as having the visual capacity to see "clearly enough to read the license plate of a car in Boston from Washington, D.C."

The optical assembly was completed in October 1984 and was shipped that November to Lockheed Missile and Space in Sunnyvale, California. Lockheed had, simultaneously with Perkin-Elmer's winning its bid to build the optical system in November 1977, received a contract from NASA to build the spacecraft that would carry the telescope into orbit—a job that was essentially complete by the end of 1983—and to install the telescope and scientific instruments into the vehicle, a task that was finished by the end of 1985.

The process, however, was far from being as smooth as this chronological summary implies. While no one ever questioned the potential of the project, its ability to be realized has frequently been the object of doubt. The telescope was almost discontinued in Congress in the mid-seventies. And then, in the early 1980s, plagued by cost overruns and production delays, the telescope's successful completion once again began to look doubtful.

An artist's conception (right) shows the space shuttle releasing the space telescope into orbit approximately 300 miles (484 kilometers) above the Earth's surface.

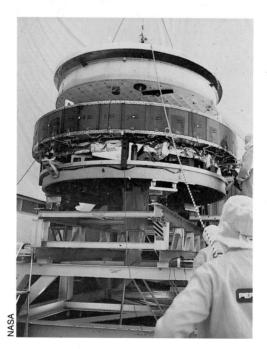

Perkin-Elmer technicians gingerly move the completed primary mirror (right) into position for mounting on the main ring, which attaches the optical assembly to the spacecraft. The mirror-to-ring mounting has been designed so that the primary mirror will be held securely in a strain-free condition while in a zero gravity environment.

NASA

NASA

BREAKDOWNS AND BREAK-THROUGHS

In August 1979, as construction was proceeding, a launch of the telescope was scheduled for sometime in 1983. That date would have to be pushed back to 1985 and then to 1986 before the telescope would be ready to go into orbit. Meanwhile, the cost of the project more than doubled. What caused these delays and cost overruns? While no one denies the technical complexity of the project, many hold the complex management structure largely responsible for its problems.

Even before Congress appropriated funds for the space telescope, NASA had divided responsibility for directing the project between its Marshall and Goddard centers. At the same time, Goddard had chosen the builders of the scientific instruments, a group of four American and two European firms. And all of the contractors, both those building the spacecraft and telescope for Marshall and those working on the instruments for Goddard, were associate contractors, meaning each was essentially equal regardless of the relative importance of its task. Then NASA divided construction of the spacecraft and optical system between Lockheed and Perkin-Elmer. Both these companies, in turn, also relied on a diverse group of subcontractors.

How did these many subcontractors feel about this system? Robert Bless of the University of Wisconsin, who has been building instruments for space-based astronomical observations since 1960 and contributed a high-speed photometer to the project, notes, "The division between Goddard and Marshall was a mistake from the beginning." Similarly, the division of responsibility between contractors Lockheed and Perkin-Elmer created problem after problem. Explains

NASA

In 1972, NASA designated the Marshall Space Flight Center in Huntsville, Alabama (shown in an aerial view on the previous page) as the lead center for the space telescope project from inception through launch. At the same time, the Goddard Space Flight Center in Greenbelt, Maryland (right) was named lead center for postlaunch operations in charge of developing the five scientific instruments that the telescope carries into space. Critics claim that this division of responsibilities led to delays and cost overruns.

Bless, "It just made getting anything more difficult, because you would have to find somebody at Perkin-Elmer who was supposed to know the answer, but he would say that it was somebody at Lockheed who knew the answer. Then, sometimes the person at Lockheed would say 'That's not our responsibility, that's Perkin-Elmer's responsibility.' "

Lockheed engineer Bertram Bulkin, a veteran of more than twenty years in the aerospace industry, agrees that along with the technical complexity of creating a space-worthy telescope the management structure was the greatest difficulty the project faced. At Lockheed, the communications with contractors, subcontractors, and various NASA centers were particularly entangled. Not only

Courtesy NASA/Goddard Space Flight Center

was Lockheed in constant, if not consistently close, contact with Perkin-Elmer, the five scientific instrument teams, and the Marshall and Goddard NASA centers, but they were also involved in instructing the astronauts about the kind of repair and replacement work the space telescope would require from the shuttle crews. Bulkin explains, "Why was it so difficult? I think the multiplicity of interfaces was the prime reason. The space telescope may well be the most sophisticated spacecraft flown so far, and dealing with all the people involved made the project extremely tough."

As a result of this diffusion of responsibility, the project began in the early 1980s to fall further and further behind schedule. Perkin-Elmer's work on the mirrors drew raves from the astronomical community, but they took much longer to manufacture than expected. And even after the mirrors were finished in December 1981, the Connecticut firm's performance was causing delays, with the result that by autumn 1982 the program was falling behind schedule at a rate of nearly one week per month.

This news came as a surprise to NASA headquarters, which, rather than keeping a close watch on the project, had preferred to believe Marshall's continuing reassurances that whatever problems had arisen were already solved. "Until things started to go to pot in the early eighties," Bless says, "NASA headquarters paid very little attention to the project." Part of the reason for this was high personnel-turnover rate at the agency. The

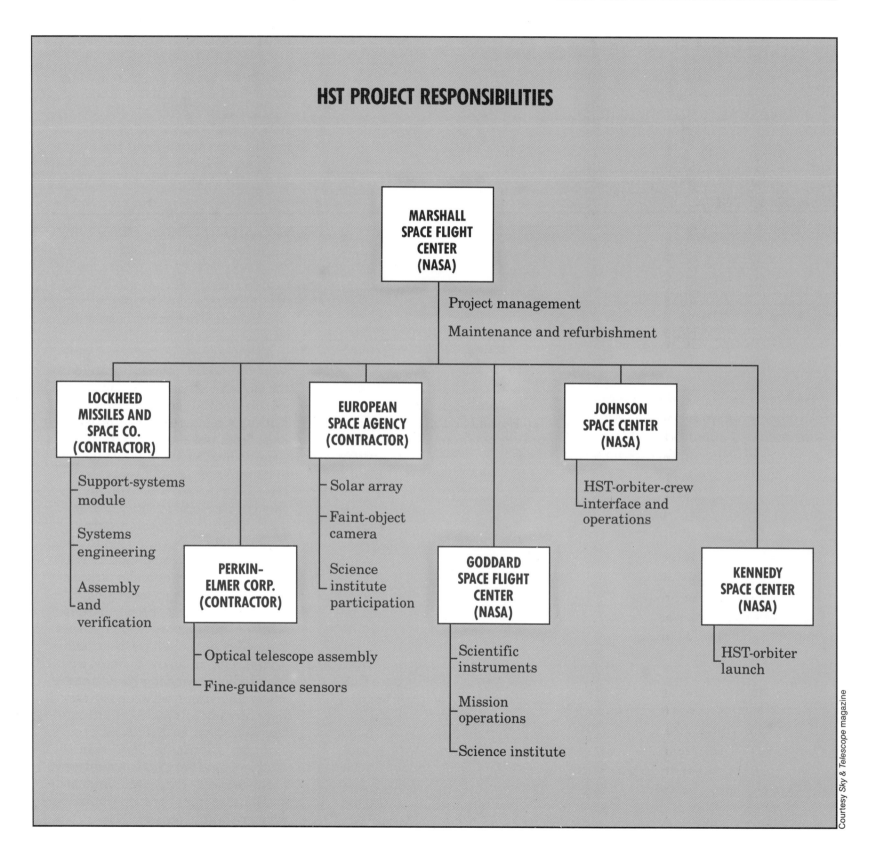

HST PROJECT RESPONSIBILITIES

MARSHALL SPACE FLIGHT CENTER (NASA)

Project management

Maintenance and refurbishment

LOCKHEED MISSILES AND SPACE CO. (CONTRACTOR)

- Support-systems module
- Systems engineering
- Assembly and verification

PERKIN-ELMER CORP. (CONTRACTOR)

- Optical telescope assembly
- Fine-guidance sensors

EUROPEAN SPACE AGENCY (CONTRACTOR)

- Solar array
- Faint-object camera
- Science institute participation

GODDARD SPACE FLIGHT CENTER (NASA)

- Scientific instruments
- Mission operations
- Science institute

JOHNSON SPACE CENTER (NASA)

- HST-orbiter-crew interface and operations

KENNEDY SPACE CENTER (NASA)

- HST-orbiter launch

Courtesy Sky & Telescope magazine

NASA

In November 1982, during a time when the space telescope project was falling behind schedule at a rate of one week per month, NASA administrator James M. Beggs (above) ordered an independent review of the program. "We underestimated the time, money, and people it would take to deal with the difficulties and uncertainties, which were inherent in the program from the beginning but went unrecognized," he later told Congress.

The complex management system NASA devised for the building of the space telescope (left) "was a mistake from the beginning," asserts University of Wisconsin astronomer Robert Bless.

NASA Office of Space Sciences and Applications had five directors during the first six years of the space telescope project.

In November 1982, NASA administrator James M. Beggs, finally aware and fully alarmed by the stalled program, ordered an independent review of the space telescope. The result: a report that termed the relationship between Marshall and Goddard "deplorable" and called for "immediate attention" to the problems at Lockheed and Perkin-Elmer. "Nobody had a tight grip on the whole project or, for that matter, any of its parts," declares a NASA official.

How had this happened? In October 1983 Beggs went before Congress to defend the program by admitting to its difficulties and arguing that they were in the process of being resolved. Beggs told the House Subcommittee on Space Science and Applications that NASA itself deserved a good deal of the blame. The space-telescope

project, he testified, is "the toughest job NASA ever tried to do." In efforts to save money, NASA had become overly "success oriented," that is, it had left little to no room in its budget for spare parts, testing equipment, and backup designs. "We underestimated the time, money, and people it would take to deal with the difficulties and uncertainties, which were inherent in the program from the beginning but went unrecognized."

And overconfidence was not the only mistake. Beggs admitted, "In addition to the technical challenge, we introduced an unusual management structure." The rivalry between Marshall and Goddard was said to have reached a point that communications between the two centers were more likely to gloss over technical glitches than attempt to solve them. To reverse these trends, Beggs appointed new management at Marshall and Goddard and began a stricter supervision of Perkin-Elmer.

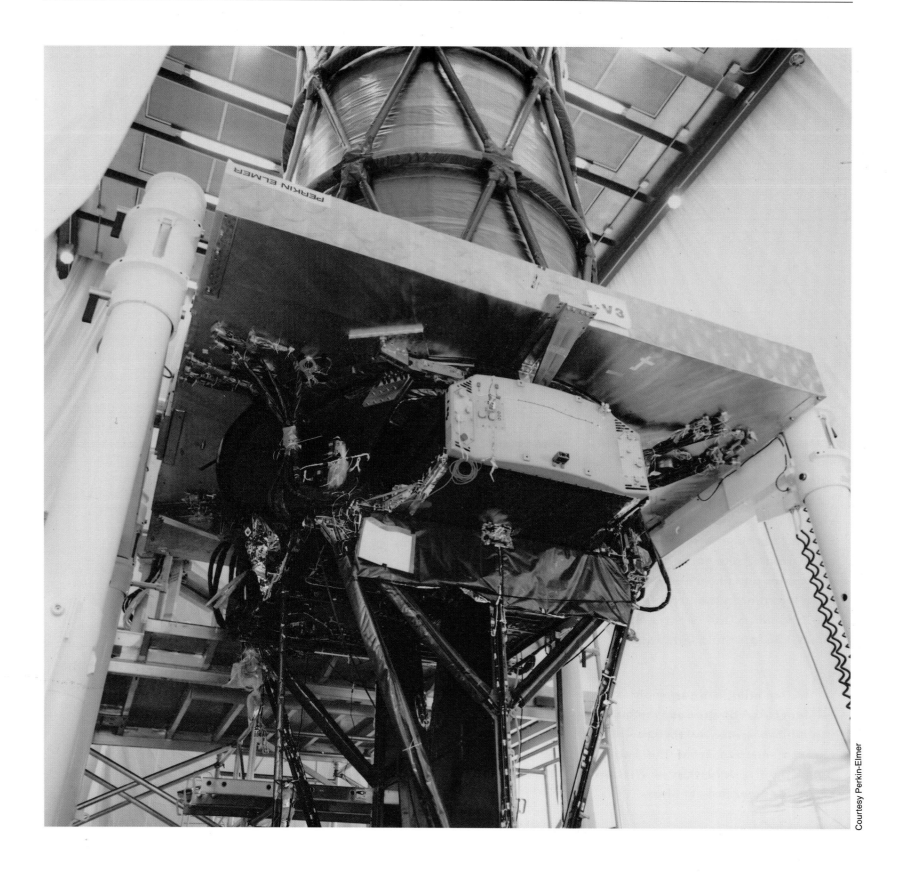

Courtesy Perkin-Elmer

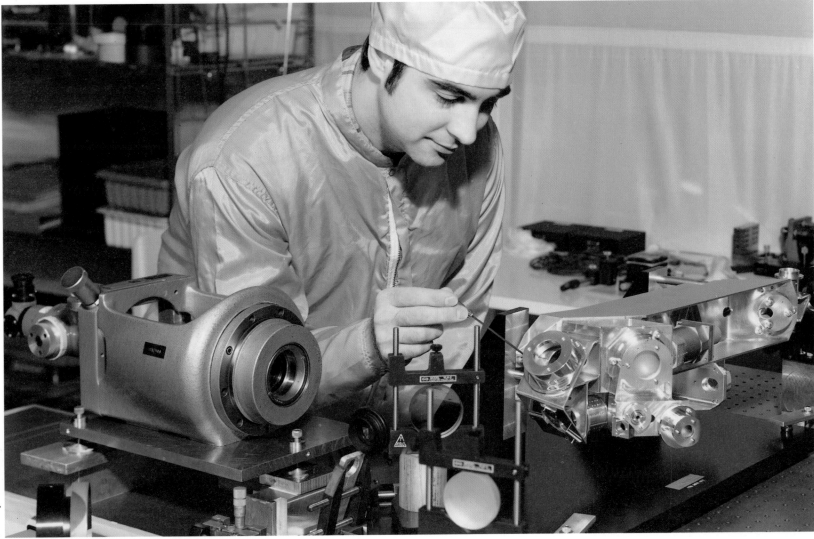

Courtesy Perkin-Elmer

The fine guidance assembly (left) caused problems because it was too fine. It was feared thát even the slightest movement of the spacecraft might cause the sensors to lose their target.

For some time technicians thought that they might have to develop a back-up system for the sensors (above), but instead they succeeded in "quieting down" the spacecraft.

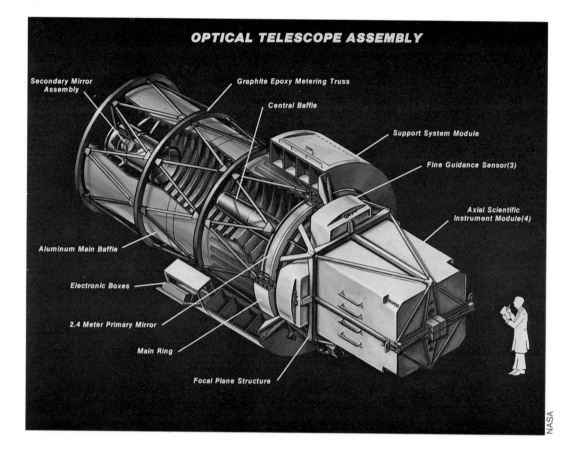

OPTICAL TELESCOPE ASSEMBLY

Secondary Mirror Assembly

Graphite Epoxy Metering Truss

Central Baffle

Support System Module

Fine Guidance Sensor(3)

Axial Scientific Instrument Module(4)

Aluminum Main Baffle

Electronic Boxes

2.4 Meter Primary Mirror

Main Ring

Focal Plane Structure

NASA

A schedule drawn up in 1979 called for the space telescope to be launched in 1983, but Perkin-Elmer did not complete the assembly of the telescope's optical system (above) until October 1984.

T he optical system arrived at Lockheed in Sunnyvale, California (right) in December 1984, and the assembly of the spacecraft, telescope, and scientific instruments was essentially complete by the end of 1985.

Courtesy Lockheed

Confused management, production delays, and cost overruns, however, are a fairly abstract way of describing the problems the project was facing. There were a couple of very concrete problems. The first was that the graphite-epoxy truss appeared likely to leak water onto and destroy the effectiveness of the telescope. Second, the fine-guidance sensors were so fine they might not work at all.

Gene Oliver was one of the people charged with confronting these very concrete issues. The son of a Mississippi oil-field contractor, Oliver grew up surrounded by cars and trucks and with the ambition of making these vehicles his life work. After graduating from Mississippi State, he joined a Chrysler Corporation program in Detroit, which offered him the opportunity to earn a masters degree in automotive engineering and a job with the company upon the completion of his studies. But in 1958, the year Oliver received his diploma, the auto industry was mired in recession and, to Oliver, Detroit did not appear an inviting place to live. Instead, he requested assignment to a program in Huntsville, Alabama, where Chrysler was assisting with the *Redstone*, an early military rocket that Werner von Braun was building for the United States Army. And when NASA absorbed the Redstone program and the von Braun team (Alan Shepard, the first American astronaut in space, rode in a Redstone rocket in 1961), Oliver left the auto firm behind to join the space agency. Oliver has been involved with the space-telescope project since the mid-

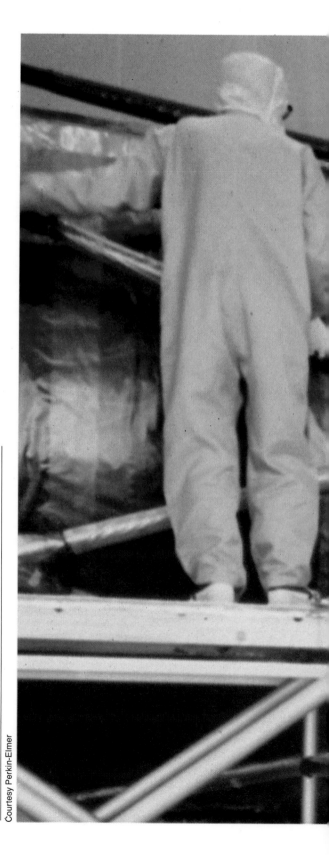

Courtesy Perkin-Elmer

The space telescope's optical system is held together by a graphite epoxy truss. Graphite epoxy is a strong, lightweight substance developed by NASA in the early 1970s that expands and contracts very little in extremes of temperature.

sixties, when the telescope was still known as the Large Space Telescope, and is one of its most dedicated supporters.

The problem with the truss, Oliver explains, is that graphite epoxy absorbs water. "It's hygroscopic," he says. "The stuff absorbs water not just on the surface but all the way in. And it reaches an equilibrium with its environment. So, because the space telescope is currently housed in a 'clean room' at Lockheed with a fifty-percent humidity, there will be a certain amount of water that stays in it. Once you get the telescope into space, the water will slowly come out of the graphite epoxy. We've already accounted for the fact that the graphite epoxy will shrink a minute amount in space, but we're concerned that the light detectors for the scientific instruments will end up covered with ice."

The solution, according to Oliver, is to subject the material to a process of purging with dry nitrogen so that as much of the moisture as possible can be eliminated before launch. Why not also try to get the humidity of the clean room down below fifty percent? "Whenever you lower the humidity in a building area where people work, static electricity becomes a problem, and there are a lot of very sensitive electronics on the space telescope," says Oliver. "We know the electrostatic capability of a room at fifty-percent humidity. If we go any lower than that, the danger caused by static electricity increases substantially."

Courtesy Lockheed

Unfortunately, graphite epoxy absorbs water from the Earth's humid atmosphere that it later releases in the humidityless atmosphere of space. Even the "clean room" at Lockheed (left), which houses the space telescope before launch, has a 50 percent humidity level.

Back to the second problem of fine-guidance sensors that might be too fine. Able to find and hold targets at distances as small as 0.007 of an arc second, at the same time they were capable of losing objects if the telescope strayed more than 0.05 of an arc second off target. "The fine-guidance sensors only work when you're very close to being on target," Oliver notes. "You have to augment them with mechanisms that work at angles as great as 30 arc seconds. The fine-guidance sensors had so many of these augmentations added that they became even more incredibly complex, and any sort of small external force was capable of causing them to lose lock." For a while, it looked like a backup pointing-and-locking system would have to be developed. But instead, the engineers succeeded in stabilizing, "quieting down" in Oliver's words, the spacecraft. "We eventually quieted the

vehicle," he says, "but in 1983 we thought we would have to work on backup systems for the fine-guidance sensors."

For a while, the discovery of new problems virtually ran neck and neck with the development of new solutions. In early 1983, it was found that the twenty-seven latches that hold the fine-guidance sensors and scientific instruments in place and allow for their possible repair or replacement by shuttle-based astronauts chafed under simulated launch conditions. Later, when Perkin-Elmer changed the coating on the latches from aluminum oxide to tungsten carbide and cobalt, the new material drastically reduced the frictional wear.

At this time, the space telescope was having so many difficulties that even problems that had been solved threatened to come unsolved. It was discovered that somehow a small part of the primary mirror, in spite of careful storage in a Perkin-Elmer clean room, was contaminated, perhaps irreversibly, by dust. Although the mirror was still "acceptable" by NASA specifications, in June 1984 Perkin-Elmer sprayed the mirror with dry nitrogen gas capable of dislodging dust particles as small as 20 to 100 microns, or micrometers, in size and then used a vacuum hose to remove them. A photographic analysis of the mirror's surface revealed that the cleaning had reduced the obscureness caused by dust from 2.4 percent to less than 0.6 percent.

The management reorganization instituted by NASA in late 1982 paid off. By January 1986, it appeared that the space telescope would be ready for its scheduled launch sometime between August and October. However, that month tragedy struck the American space program, when the space-shuttle *Challenger* exploded minutes after launching, killing the seven astronauts aboard and putting all further shuttle missions on hold while a presidential commission investigated why the disaster happened and how it could be prevented from happening again.

While the space-telescope project benefits from the increased amount of time available for testing and launch and flight simulations, no one involved is pleased by the delay, and some are quite worried. Perkin-Elmer physicist Facey, recalling the dust that had settled on the mirror in that company's clean room, says that the telescope would be a lot safer up in space. "The worst place you could keep it," he muses, "is on the ground."

NASA

A computer drawn photograph (above) of a top view of the *Challenger* suggests that a solid rocket booster malfunction was responsible for the disaster.

NASA

The launch of the space telescope was scheduled for the autumn of 1986, but was postponed indefinitely after the space shuttle *Challenger* (left) blew up 73 seconds after liftoff on January 28 of that year.

FEASTING ON STARLIGHT

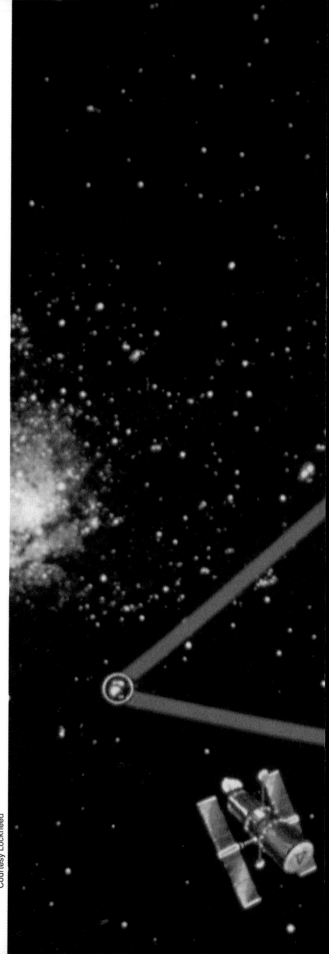

Courtesy Lockheed

Growing up in Illinois farm country, outside the town of Salem about 60 miles due east from St. Louis, David Leckrone had a splendid view of the night sky and, having already decided at the age of nine to become a scientist, he made the most of it. "I just enjoyed looking at the sky," he recalls. "I was just fascinated by it, and I did things no sane person would do today, like make smoked glass in order to look at sunspots. That can be pretty bad for your eyes. I looked at comets and at meteor showers. I even saw the aurora borealis—the northern lights—a few times as a boy. I had a natural affinity for the subject, and small towns in Illinois can be very boring. There's not that much else to do."

Smoked glass was not the only optical instrument Leckrone experimented with. "I kept wanting to get a telescope so I could really

see," he continues. "It was frustrating. I tried binoculars, and then I had a very small telescope, but I could never see what I wanted to see. Finally, when I was a senior in high school, I built a 4-inch reflector. For the first time, I actually saw globular clusters and the Orion nebula, and that really turned me on."

Since 1976, astronomer Leckrone has been involved in designing and building the most complex, most expensive and, in the hopes of its creators, most effective optical telescope ever made. His specific task, as the Goddard Center's space telescope instruments scientist, has been to supervise the selection and coordinate the construction of the five scientific instruments the space telescope will carry when it is launched.

The instruments commissioned by Goddard are a wide-field/plane-

Orbiting the Earth at a height of 300 miles (482 kilometers), a distance freeing it from most of the distorting effects of the atmosphere, the space telescope (previous page) will be able to resolve objects ten times better than ground-based observers.

tary camera (that is, a camera with distinct settings, one for photographing relatively large sections of the sky, the other for taking detailed images of relatively close objects, like planets), a faint-object camera, a high-resolution spectrograph, a faint-object spectrograph, and a high-speed photometer. Orbiting 300 miles (482 kilometers) above Earth, high enough to avoid most of the screening and distorting effects of the atmosphere, these five instruments along with the fine-guidance sensors will feast on starlight gathered by the telescope's optical system and record their observations on computer tape for ground-based astronomers to analyze. And ground-based astronomers will also be aiming the telescope and instruments at their targets. Therefore, a continuous

NASA

Astronauts from the space shuttle (left) will visit the space telescope periodically to perform maintenance, and even to upgrade the scientific instruments.

human presence will guide the telescope although no one will actually be aboard. It will be placed in orbit by shuttle astronauts, who may also return occasionally for maintenance visits.

While the absence of a human presence on board may seem odd to the layman, astronomers long ago abandoned such direct observation. "The traditional caricature of an astronomer has always been a white-haired, white-bearded old man peering myopically through his long telescope tube," comments British astronomer Nigel Henbest. "For the past century, however, astronomers have only rarely looked through their telescopes. The human eye is simply not a very good light detector."

Back in the 1960s, when the space telescope was known as the Large Space Telescope, the potential of unmanned observatories, while obvious in the astronomical community, had not entirely captured the imagination of nonscientists. "At the time we started promoting the space telescope," notes Spitzer, "our chief argument with many people in NASA, and with space enthusiasts inside and outside the government, was that we felt the telescope should not be operated by astronauts up in the spacecraft with it. We had quite a battle to convince people that the proper role of man with an instrument of this sort was not routine operation but occasional maintenance and updating."

In 1976, Leckrone helped to draft a notice, officially called the "announcement of opportunity," which was sent out by Goddard to the scientific community to solicit proposals for the telescope's scientific instruments. "We left the announcement-of-opportunity process open because of the possibility that there were going to be new ideas and new developments in instrumentation that would come out of the woodwork and be more interesting than what had been defined in the preliminary studies," he says. In 1977, Goddard received a dozen or so proposals.

While the Goddard scientists attempted, as Leckrone put it, "to select the best instruments based on the best proposals," NASA had made it clear to them that based on the preliminary studies there were two specific instruments, a wide-field/planetary camera and a faint-object spectrograph, that would be necessary to justify the effort and expense of sending the telescope into orbit. Because the camera is expected to take visually stunning pictures (rather than readings of spectra, images of faint objects, or high-speed light readings), it will probably be the instrument that will garner the bulk of the popular and media attention.

But, although the camera will probably have the largest impact on the public, spectrographs are the basic tools of modern astrophysics. And the faint-object spectrograph will enable astronomers to investigate objects difficult or even impossible to observe with ground-based telescopes. Consequently, the camera and spectrograph were identified in the announcement of opportunity as being singularly important categories of instrumentation, with the predictable result

A diagram (right) illustrates the space telescope with the five scientific instruments—including two cameras, two spectrographs, and a photometer—that it will carry into orbit to feast on the light collected by its primary mirror. An artist's conception (above) provides a stunning view of the telescope in orbit.

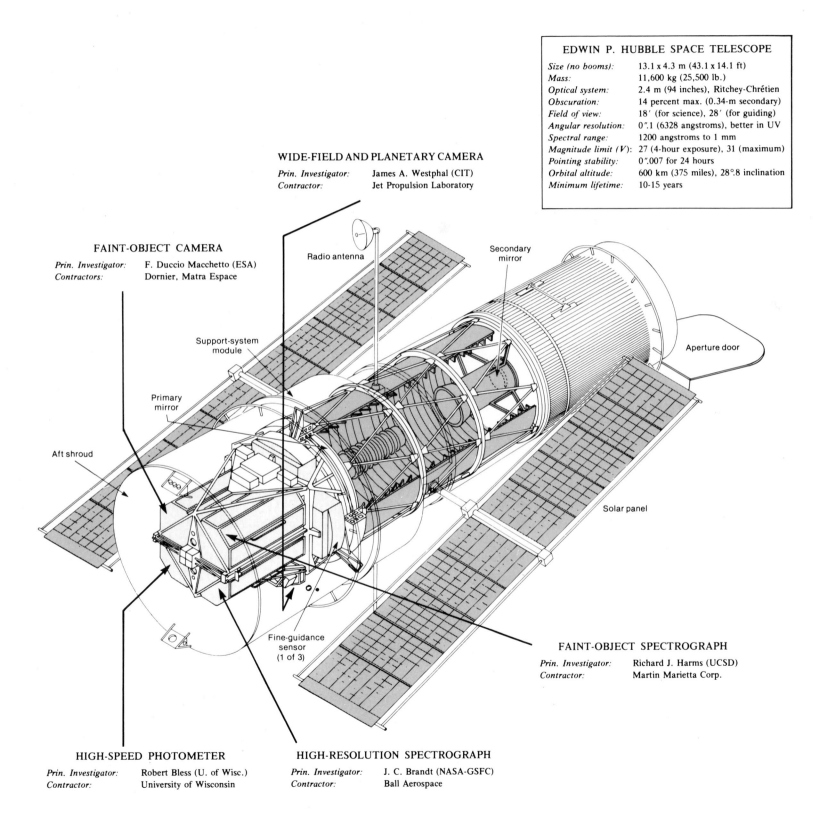

EDWIN P. HUBBLE SPACE TELESCOPE

Size (no booms):	13.1 x 4.3 m (43.1 x 14.1 ft)
Mass:	11,600 kg (25,500 lb.)
Optical system:	2.4 m (94 inches), Ritchey-Chrétien
Obscuration:	14 percent max. (0.34-m secondary)
Field of view:	18′ (for science), 28′ (for guiding)
Angular resolution:	0″.1 (6328 angstroms), better in UV
Spectral range:	1200 angstroms to 1 mm
Magnitude limit (V):	27 (4-hour exposure), 31 (maximum)
Pointing stability:	0″.007 for 24 hours
Orbital altitude:	600 km (375 miles), 28°.8 inclination
Minimum lifetime:	10-15 years

WIDE-FIELD AND PLANETARY CAMERA

Prin. Investigator:	James A. Westphal (CIT)
Contractor:	Jet Propulsion Laboratory

FAINT-OBJECT CAMERA

Prin. Investigator:	F. Duccio Macchetto (ESA)
Contractors:	Dornier, Matra Espace

Radio antenna

Secondary mirror

Support-system module

Primary mirror

Aft shroud

Aperture door

Solar panel

Fine-guidance sensor (1 of 3)

FAINT-OBJECT SPECTROGRAPH

Prin. Investigator:	Richard J. Harms (UCSD)
Contractor:	Martin Marietta Corp.

HIGH-SPEED PHOTOMETER

Prin. Investigator:	Robert Bless (U. of Wisc.)
Contractor:	University of Wisconsin

HIGH-RESOLUTION SPECTROGRAPH

Prin. Investigator:	J. C. Brandt (NASA-GSFC)
Contractor:	Ball Aerospace

Courtesy Sky & Telescope magazine

that the competition to produce them was especially intense, particulary in the case of the wide-field/planetary camera.

As its name implies, the wide-field/planetary camera is the most versatile of the space telescope's scientific instruments. It has two essential functions: As a wide-field camera it can be used to take pictures of large sections of the sky, and as a planetary camera it can take high-resolution images of nearby objects such as planets. In some cases, the images would be of about the same quality as those taken by flyby planetary missions.

In 1977, Goddard received two major proposals for the camera: one from a Princeton group headed by Spitzer, which suggested using a then standard television tube detector; the other from a team at the California Institute of Technology with the Jet Propulsion Laboratory (JPL) in Pasadena as a subcontractor to do the actual manufacturing, which recommended employing a new technical development, a charged coupling device (CCD). "If the proposal evaluation had taken place six months earlier, we in all likelihood would have gone with the Princeton concept," says Leckrone. But instead, Goddard decided to take a chance.

CCDs have greatly aided astronomers in the recent past. Its inventors at Bell Labs, however, had originally hoped to put the devices to a more domestically oriented use. One of the principal attractions of the Bell Telephone exhibit at the 1964–1965 New York World's Fair was a Picturephone, which permitted users to establish both

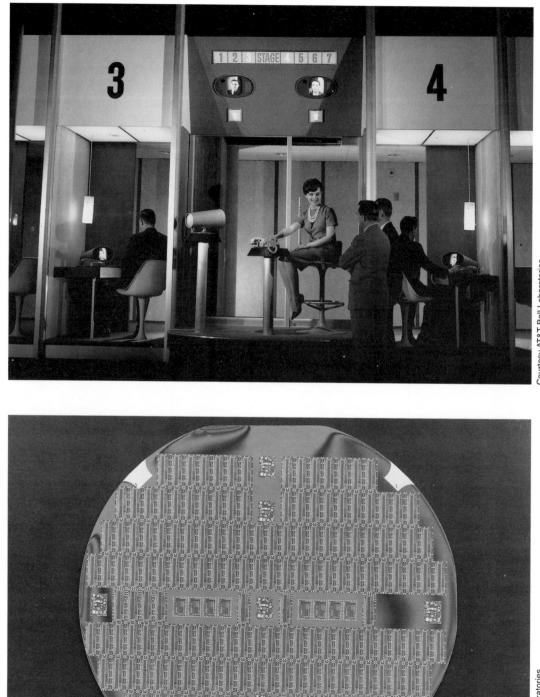

Courtesy AT&T Bell Laboratories

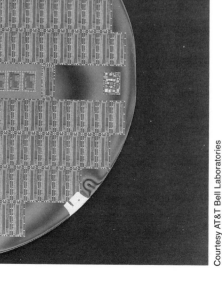

Courtesy AT&T Bell Laboratories

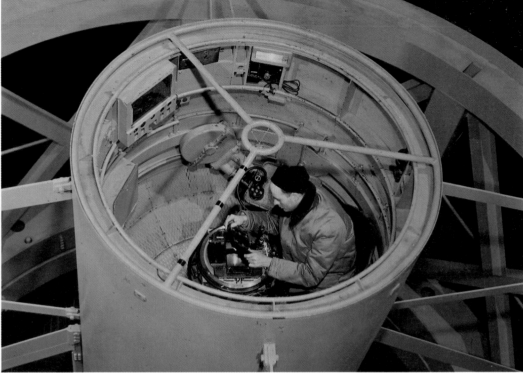

Courtesy The Carnegie Institution of Washington

Charged coupling devices (CCDs) (below left), the light detectors on the space telescope's wide-field planetary camera, were a product of Bell Laboratories's unsuccessful efforts in the 1960s to produce a practical Picturephone (above left), a version of which was a popular exhibit at the 1964–1965 New York World's Fair. CCDs are several hundred times more sensitive to light than the photographic plates in use when the Mount Palomar telescope (above) went into operation in 1948.

audio and visual contact with each other. At the fair, such calls could take place only between paired booths within the Bell Pavilion, but Bell's ultimate intention was to transform this novelty into a household object. After all, the previous New York World's Fair, in 1939 to 1940, had introduced thousands of people to television, which at that time had seemed an unlikely candidate to supplant radio from its central place in the American home.

But, try as they might, the inventors at Bell Labs did not succeed in producing a phone that would enable a caller to reach out and see someone. In their efforts to develop a suitable picture element, however, they did invent CCDs. A CCD is essentially an oversized silicon chip that converts light that falls on it into a sequence of electric signals. In 1977, astronomers at the California Institute of Technology got the idea of using the new devices as light detectors on a space-telescope science instrument. They took a CCD down to Mount Palomar, used it on the 200-inch (5-meter) telescope there, and were favorably impressed. There was one remaining problem with the detectors, though. They did not work in the ultraviolet range, the type of radiation above the Earth's interfering atmosphere that the telescope was expected to explore in detail. The Cal Tech team solved this problem by coating half of each chip with coronene, a phosphor that converts ultraviolet light to a yellow-green to which the devices are extremely sensitive.

Each CCD used in space telescope is a square measuring 0.5

SCIENTIFIC CAPABILITIES
- PLANETS, STARS, GALAXIES, DIFFUSE MATTER
- ULTRAVIOLET, VISIBLE AND INFRARED
- DETECT OBJECTS 7 TIMES FAINTER THAN PRESENT
- OBSERVABLE VOLUME OF SPACE MORE THAN ONE-HUNDRED TIMES GREATER THAN PRESENT
- 7–10 TIMES BETTER RESOLUTION THAN GROUND-BASED TELESCOPES
- IMAGING, PHOTOMETRY, SPECTROSCOPY

CHARACTERISTICS
- 2.4 METER DIAMETER MIRROR
- RITCHEY–CHRETIEN TELESCOPE
- ACCOMODATES UP TO 5 SCIENTIFIC INSTRUMENTS
- WEIGHT–20,000 LBS.
- SHUTTLE LAUNCHED
- IN-ORBIT MAINTENANCE AND UPDATE
- RECOVERY FOR REFURBISHMENT AND RELAUNCH
- LIFETIME—GREATER THAN DECADE
- INTERNATIONAL PARTICIPATION

NASA

An artist's rendering of the space telescope (left) lists its remarkable capabilities. The competition among astronomers to design the scientific instruments led several of them to push available technologies to new limits.

inch (1.2 centimeters) on a side that contains 640,000 picture elements, or pixels, arranged in 800 columns with 800 pixels per column. CCDs allow astronomers to collect light with a spectacular efficiency. When it was completed in 1948, Mount Palomar's telescope used photographic plates that registered only 0.0033 of the light that struck them. The best photographic plates developed since then can record about 0.033 of the light that falls on them. The televisionlike image tubes, the technology that preceded CCDs, could respond to about 0.25 of the available light. CCDs can detect 0.75 of the light that strikes them.

While the images from this camera earn their glory in the imaginations of nonscientists, astronomers will direct at least as much atten-

tion to the readings from the space telescope's spectrographs. Several years after Herschel accidentally discovered the existence of infrared radiation through the use of a prism, a German optician, Joseph von Fraunhofer, set the stage for the use of the prism as an effective tool to analyze light. Von Fraunhofer found that if only a narrow slit of light is allowed to pass through the prism, as in the instrument we now call a spectroscope, a series of dark lines appear in the spectrum. Then, around 1860, a pair of German chemists, Wilhelm Bunsen (who had invented that ubiquitous piece of laboratory equipment, the Bunsen burner, five years earlier) and Gustav Kirchhoff, discovered that when one examines the glow from a variety of burning substances, each substance produces a

characteristic set of spectral lines.

Today, astronomers use spectrographs to analyze the chemical composition of celestial objects and, thanks to additional discoveries about the nature of spectra, to determine the temperature and movement of objects as well. The space telescope will carry two spectrographs into orbit, but, in contrast to the situation with the wide-field/planetary camera, new detector technology did not arrive on the scene early enough to be incorporated in either of them.

Albert Bogges, who served as project scientist and chief designer of the *International Ultraviolet Explorer*, has, in his words, "worn several space-telescope hats," over the years. He was first the head of a group that researched spectrographs in the preliminary design

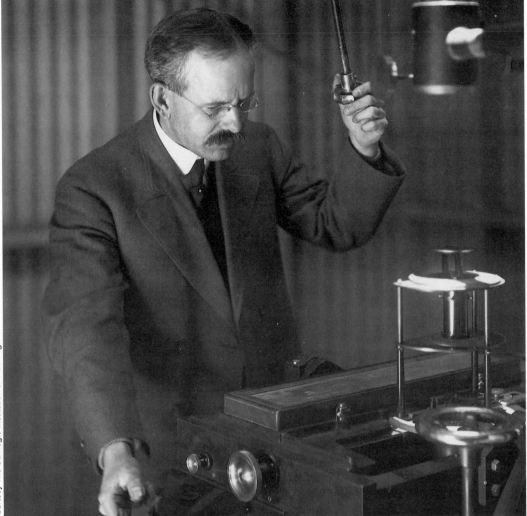

Courtesy The Carnegie Institution of Washington

Astronomer George Ellery Hale (above) played key roles in the building of several of the twentieth century's largest and most important telescopes: the University of Chicago's Yerkes refractor, the 100-inch (2.5-meter) reflector at Mount Wilson, and the 200-inch (5-meter) reflector at Mount Palomar.

phase; was then codeveloper with his Goddard colleague John Brandt of the space telescope's high-resolution spectrograph; and since 1983 has been a space-telescope project scientist at Goddard.

"The most efficient and effective way to build a modern spectrograph is to use a detector that can create an image, something not all detectors do," explains Bogges. "What you want a spectrograph to do is measure brightness at each of many different wavelengths, and the very simplest of spectrographs have a one-element detector that just looks at one wavelength at a time before moving, physically or using a mirror, to the next. This is the simplest but also the least efficient of methods. It would be much more efficient if you could look at all the data you wanted simultaneously by forming an image."

That would be possible, according to Bogges, using some sort of television system. "When we were first laying out the concepts for the spectrograph, we assumed a television system would be available," he says. "NASA had spent a considerable amount of money helping to develop television cameras for use in astronomy, and that was what all our original plans were based on. But, by the late seventies, when it came time for people to make concrete proposals for instruments that they would actually build and stake their reputations on, it was apparent that television cameras were not going to do the job as well as we had all hoped and expected."

One possible alternative was the experimental technology of CCDs. Bogges and his fellow spectrograph designers, however, decided to take a more conservative approach. "In this business, when you make a proposal, you not only have to think of something that you know will work but also something that will be selected," he observes. "We thought it was important to go with something whose effectiveness was easily demonstrable, so we chose a Digicon, a device that was already in use on ground-based telescopes at the time." A Digicon sensor, operating on the basis of the photoelectric effect, converts the incoming light into photoelectrons, which it proceeds to arrange in a single line that is 512 units long, with each unit representing a particular wavelength.

The faint-object spectrograph—which, along with the wide-field planetary camera, was one of the two instruments given special priority by NASA in its preliminary studies—was developed by a team of scientists at the University of California at San Diego and built by the Martin Marietta Corporation of Denver, Colorado. This spectroscope uses a pair of photon-counting Digicon detectors, one of which is sensitive to red light and the other to blue light and ultraviolet radiation. A special feature of this spectrograph is its ability to register rapid variations in light, perhaps as fast as a few milliseconds.

The other spectrograph aboard the space telescope, the high-resolution spectrograph, developed at Goddard by Bogges and Brandt and manufactured by Ball Aerospace of Boulder, Colorado, uses a single Digicon sensor to examine light that

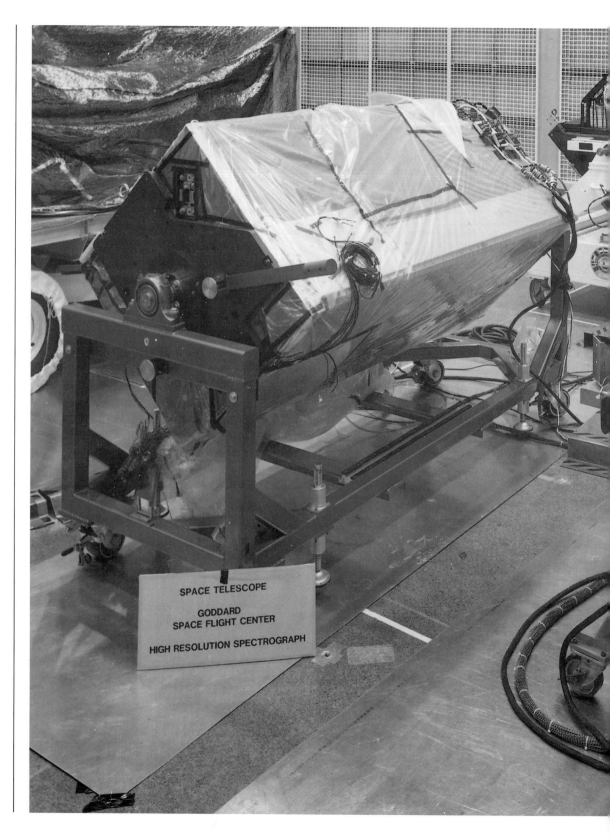

SPACE TELESCOPE

GODDARD
SPACE FLIGHT CENTER

HIGH RESOLUTION SPECTROGRAPH

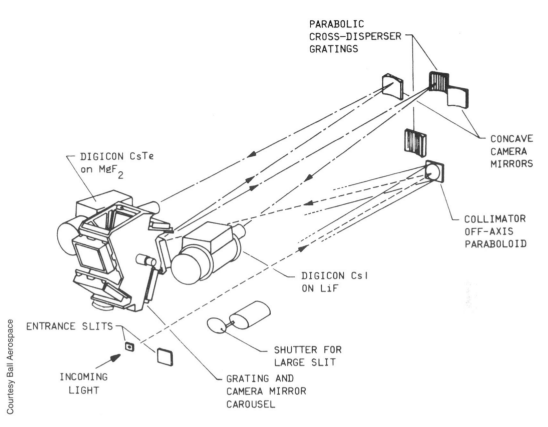

PARABOLIC CROSS-DISPERSER GRATINGS

CONCAVE CAMERA MIRRORS

DIGICON CsTe on MgF$_2$

COLLIMATOR OFF-AXIS PARABOLOID

DIGICON CsI ON LiF

ENTRANCE SLITS

INCOMING LIGHT

SHUTTER FOR LARGE SLIT

GRATING AND CAMERA MIRROR CAROUSEL

Courtesy Ball Aerospace

SPACE TELESCOPE FSC SCIENTIFIC INSTRUMENTS CONTROL AND DATA HANDLING SYSTEM

Courtesy NASA/Goddard Space Flight Center

The diagram (above) and the photo (left) show the High Resoluton Spectrograph (HRS), one of five scientific instruments that share the focal flame of the 100-inch (2.5 meter) diameter space telescope. The instrument has six gratings and an acquisition mirror on a rotatable carousel that provides three types of resolution. Ball Aerospace Systems Division (BASD) designed, fabricated, calibrated, and tested the proto-flight model of the HRS. Integration of the instrument into the space telescope is currently underway.

Courtesy NASA/Goddard Space Flight Center

SPACE TELESCOPE
EUROPEAN
SPACE AGENCY
FAINT-OBJECT
CAMERA

Protected by special suits, hats, and gloves, the highly skilled technicians (left) of the European Space Agency proudly stand beside the faint-object camera of the space telescope. The extremely sophisticated instrument will help in observing the nature and source of faint astronomical objects. The technicians (right) service the HRS, an extremely refined instrument that will be capable of examining the ultra-violet aspects of spectra and light that do not reach the Earth.

has passed through one of the six interchangeable diffraction gratings mounted on a rotating carousel, the spectrograph's only moving part. This spectrograph will produce a resolution twenty to one hundred times as good as that of the faint-object one, but will only be able to do this for stars that are some sixty to three hundred times brighter than those the faint-object spectrograph can observe. Working exclusively in the ultraviolet region of the spectrum, the high-resolution spectrograph will follow in the path begun by *Copernicus* and the *International Ultraviolet Explorer* but will do so with considerably greater resolution. It will be able to see stars as faint as the thirteenth magnitude, six magnitudes fainter than *Copernicus* and four fainter than *IUE*, and with

considerably better resolution than either of its precursors.

Although the American media still tend to portray space as the exclusive domain of the United States and the Soviet Union, the ESA has in a little more than a decade established Western Europe as a solid presence in extraterrestrial efforts. Formed in 1975 to consolidate the activities of the European Research Organization and the European Launcher Development Organization, the ESA claims thirteen member states: Austria, Belgium, Denmark, France, West Germany, Ireland, Italy, the Netherlands, Norway, Spain, Sweden, Switzerland, and the United Kingdom. In a previous collaboration with NASA, the European agency built *Spacelab*, the manned space laboratory first flown as part of a

November–December 1983 space-shuttle mission and subsequently used three more times in 1985. Without American participation, ESA flew *Giotto*, its first interplanetary project, which carried ten scientific instruments within several hundred miles of Halley's comet and produced some spectacular images of what was, to the earth-bound, the unspectacular and extremely disappointing return of the celebrated celestial body.

The ESA developed a faint-object camera for the space telescope, with Dornier Systems of West Germany and British Aerospace doing the actual manufacturing. Perhaps more than any of the other instruments, this camera exploits the full possibilities of the telescope to observe objects that are inaccessible to ground-based instruments. Actu-

Courtesy Space Telescope Science Institute

ally, it is a combination of two independent camera systems in a single package, each using a vidicon detector, an image-intensifying device similar to the light-sensitive cathode ray inside a television camera.

On one of its settings, the faint-object camera will be able to view a field that is 22 arc seconds square, about the size of Saturn viewed at opposition from Earth. This system has a series of fourteen filters and prisms as well as the capability of using an additional grating that will allow it to take spectroscopic readings of a target object. The other system has four filter wheels that contain forty-four different prisms, filters, and polarizers. The system also has two coronographic "fingers" that permit it to screen out a bright object that might obscure a nearby faint target. Its highest possible resolution, which will be used for ultraviolet studies, will create a field of view only 4 arc seconds on each side that will allow astronomers to see cloud features only a few miles across on Jupiter.

The simplest instrument on the space telescope, the high-speed photometer developed and manufactured at the University of Wisconsin by a team headed by Bless, has no moving parts. Depending completely on the telescope's fine-pointing system to direct light onto one of its roughly one hundred combinations of filters and apertures, this instrument is capable of distinguishing such events as changes in the brightness of stars only 10 microseconds apart. Cathode-ray tube technology is a tradition at the University of Wisconsin dating back to the 1920s, one which the

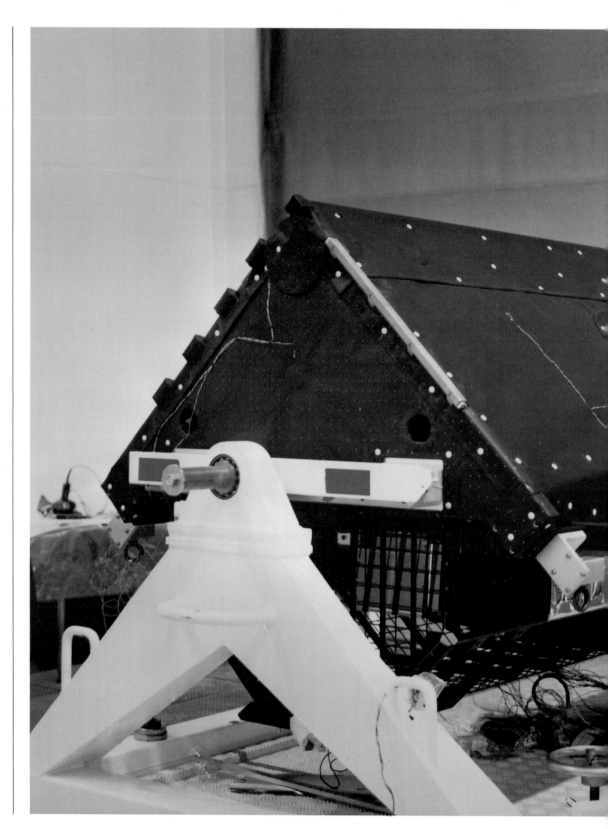

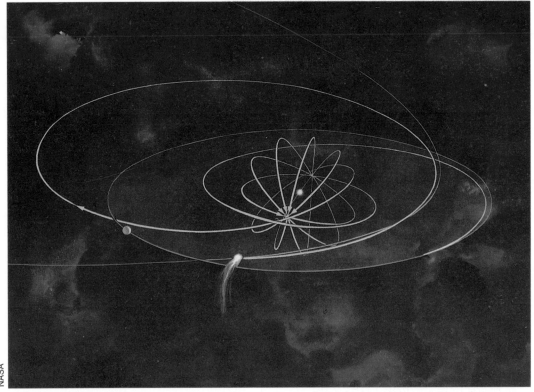

NASA

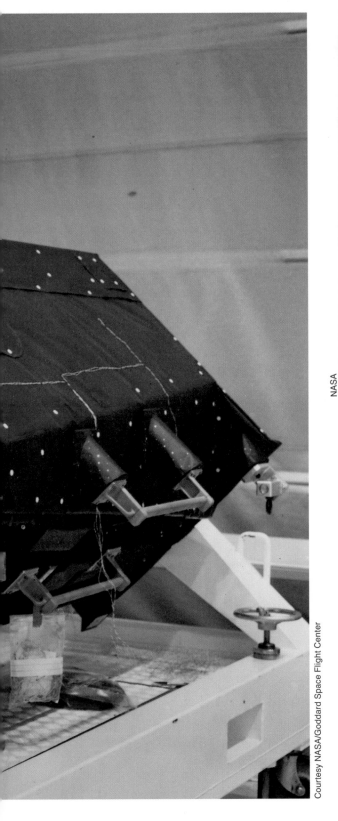

Courtesy NASA/Goddard Space Flight Center

An artist's rendering (above) portrays the legendary Halley's Comet as its orbit approached the Earth in 1986. The comet was previously sighted in 1910.

The encased HRS (left) uses light and resolves it into fine increments making it an extremely accurate instrument. The spectrograph will help in the study of such objects as supernovas, galaxies, quasars, and more, detecting images 1,000 times dimmer than previously observed.

high-speed photometer will carry into orbit. In the earliest cathode-ray detectors, a beam of light would strike the tube to be converted at the other end into a stream of electricity. After World War II, the photomultiplier tube came in use. As its name implies, this device could multiply, as much as a million times, a weak input of light into a powerful output of electricity. The detectors the photometer will use, called image dissectors, resemble photomultipliers with the exception that after the light is converted into electricity and the electricity multiplied the dissectors can respond to the photoelectrons coming from a very small region of the cathode.

These are the five instruments that the space telescope carries at launch. They are not, however, the only instruments that will be aboard the telescope during its projected fifteen-year lifespan. An essential aspect of the project, as it was conceived in the mid-seventies,

is its ongoing relationship to the space shuttle. The shuttle carries the telescope into space and releases it into orbit, then future shuttles will carry astronauts who will revisit the telescope and repair or replace malfunctioning parts. However, not all parts will be replaceable. If the primary or secondary mirrors—perhaps the telescope's most important parts—develop flaws, the best that can be hoped is that the astronauts will be able to reload the telescope aboard their spacecraft and bring it back to Earth.

Still, each of these five scientific instruments is designed to be replaceable in orbit, as are the telescope's flight computers, batteries, various electronic systems, and solar arrays, among other components. What remains unclear is how often the shuttle will be able to visit the telescope. "We don't want to touch the spacecraft as long as it is working well," explains Leckrone. "We will be making a

Pictured here is the night sky as it appeared from Las Campanas Mountain in Chile on February 22, 1987 (above), and February 23, 1987 (right), the day before, and the same day that Canadian astronomer Ian Shelton discovered the brightest supernova (arrow) since 1604. Although the space telescope missed this cosmic explosion, it will provide astronomers with a superb view of many unforeseen celestial events in the future.

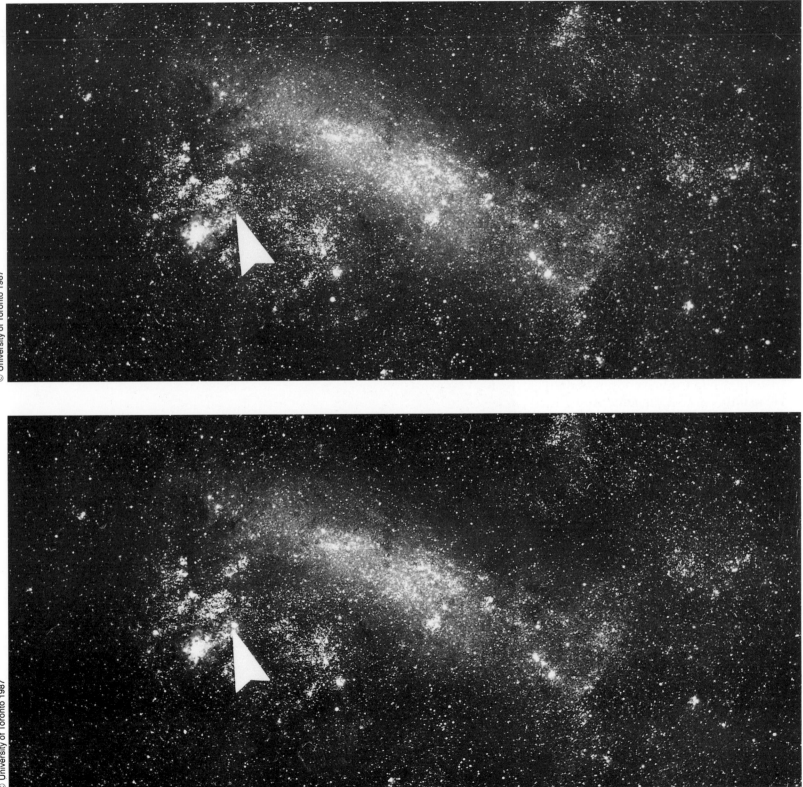

© University of Toronto 1987

© University of Toronto 1987

regular series of decisions about whether to visit the telescope rather than necessarily making a regular series of visits. On the other hand, if something goes wrong, we can't just hop up in the shuttle and fix it. The fastest that could be arranged, in a true emergency, a real crisis, would be from six to nine months." And how long are the instruments expected to last? "They're designed to function for from two and a half to three years," answers Leckrone.

Despite a relatively short expected lifespan not every one of the replaceable parts will have a backup. "The sad thing is that there's not enough money in the NASA budget—or maybe in the whole world—to build enough new instruments," observes Leckrone. Instead, Goddard has chosen to develop only three replacement instruments, and they will be staggered in their development time so that all three are not ready at once. The first is essentially a substitute wide-field/planetary camera, so close to the original that scientists refer to it as the wide-field clone.

The second is a spectrograph that combines the functions of the two spectrographs the space telescope carries into orbit upon its initial launch. The design of astronomical instruments, like the design of personal computers, is a rapidly advancing field. And if a five-year-old computer is likely to be obsolete, a ten-year-old spectrograph (the plans for the original five instruments were submitted in 1976) is already an antique. And the replacement spectrograph is not only more versatile than its

predecessors, it is also a superior high-resolution and faint-object device.

The third replacement instrument neither replicates nor improves upon the characteristics of any of the original instruments. Instead, it will add a new capability, that of observing objects in the near infrared. Leckrone notes, "It's not clear if three replacement instruments will be enough to assure that there will always be five working instruments aboard space telescope at any one time."

The experience of recent scientific space missions, however, suggests that the instruments may last longer than their designers expect. The *IUE*, for example, considerably exceeded all reasonable expectations. "It is typical for a well-designed satellite to last longer than expected, simply because good engineers are conservative about their designs," comments Leckrone's Goddard colleague Bogges. "The basic design for *IUE* was done in the early seventies. At that time, NASA was in the habit of building satellites that had intended or designed lifetimes of several months up to a year. It was our contention that experience showed that these satellites in fact lasted much longer. We as scientists proposed to build *IUE* with an intended lifetime of five years. The engineers were just frightened to death of the prospect of having to guarantee a five-year lifetime. After considerable negotiation, we compromised on a three-year lifetime. By now, the *IUE* has completed more than nine years in orbit."

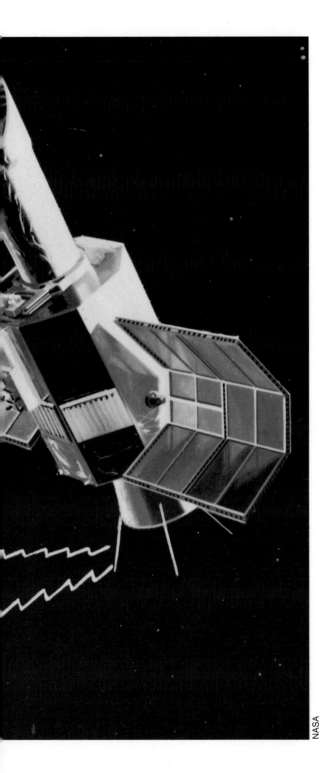

NASA

The International Ultraviolet Explorer (IUE), one of the most successful of NASA's astronomical satellites, exceeded its expected three-year lifetime by many years.

INSIDE THE OBSERVATORY

The astronomers who perform observations on the space telescope have it easy. Not only do they not have to venture into space, they don't even have to travel to a Chilean mountaintop or the crater of a Hawaiian volcano and work in a perpetually cold, oxygen-scarce environment as do other astronomers. Instead, they only have to go to Baltimore, Maryland, where a new building in a bucolic corner of Johns Hopkins University houses the Space Telescope Science Institute.

Readings from the space telescope's science instruments are transmitted to one of two satellites in NASA's Tracking and Data Relay Satellite System (TDRSS). These satellites are in geosynchronous orbit. They go around Earth at the same speed at which the planet rotates and, therefore, are always above the same location on the ground at an altitude of 22,500 miles (36,200 kilometers). The TDRSS satellite that the telescope contacts then beams the data down to its receiving station in White Sands, New Mexico, which then beams the data back up the same distance to a commercial communications satellite. This satellite transmits the data to the Goddard Center. From Goddard, land lines carry the information the final 40 miles (64 kilometers) to the Johns Hopkins campus.

How did a project in which the diffusion of responsibility, as many critics maintain, had already led to a cumbersome and inefficient management system find itself with yet another component? Astronomers themselves had been calling for a science institute for the space telescope since the project was first being given serious consideration in the early 1960s. Optical astrono-

NASA

The satellites in NASA's Tracking and Data Relay Satellite System (TDRSS) (previous page) orbit the Earth at the same speed as the planet, and, therefore, are always above the same location on the ground. The space telescope will relay its observations to one of two TDRSS satellites, which will then beam the data down to its receiving station in White Sands, New Mexico, as shown in an artist's conception (right).

mers, in particular, enjoyed a reputation for independence and were used to running their observatories themselves through university consortia with little or no government interference. In addition, astronomers in general had few dealings with NASA and, while impressed by its engineering achievements, were skeptical of its ability to do science. (More recently, NASA, and especially its Goddard Center, have more than demonstrated their skill at handling such spectacularly successful scientific missions as the *IUE* and the *Einstein* X-ray satellite.)

By the early 1970s, when NASA began preliminary studies for the space telescope, the project's backers in the astronomical community—people like Spitzer and even O'Dell, NASA's own project

scientist—were growing ever more insistent in their demands for a separate science institute. Responding to this pressure, in 1976 NASA asked the National Academy of Sciences to form an ad-hoc committee of astronomers to consider the issue. The committee considered a separate science institute a fine idea in its report. NASA accepted the report and proceeded to stall. Not until 1980 did NASA send out a request for proposals for managing the science institute.

The competition was at least as stiff as that to build the spacecraft, optical system, or scientific instruments. Although modern telecommunications made it possible for the institute to be located anywhere, the potential contractors felt that getting as close to Goddard as possible presented advan-

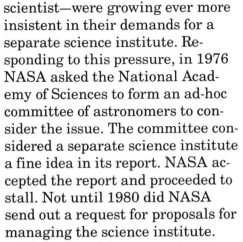

Communications from the TDRSS satellites are received at ground stations by three antennae approximately 59 feet (18 meters) in diameter like the one pictured (left).

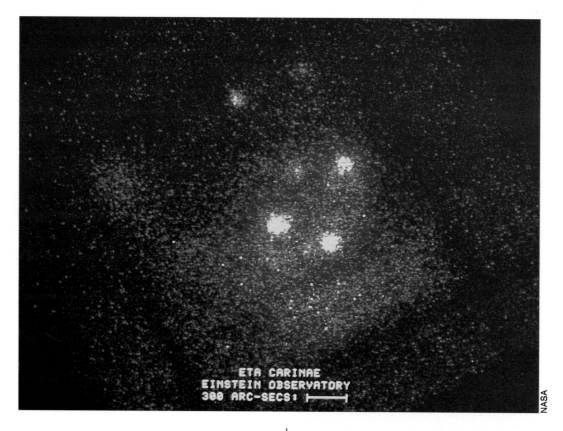

ETA CARINAE
EINSTEIN OBSERVATORY
300 ARC-SECS: |———|

NASA

Pictured (above) is an X-ray picture of several young stars. Although X-ray observations of the Sun were made in the 1940s and 1950s, most astronomers were not interested in the field until a team lead by Richard Giacconi, now head of the Space Telescope Science Institute, discovered X-ray sources outside the Solar System in 1962.

tages in terms of staffing and the creation of a high-speed data link. The sixteen-member Association of Universities for Research in Astronomy (AURA), which runs the Kitt Peak National Observatory in Arizona and the Cerro Tololo Inter-American Observatory in the Chilean Andes, approached the University of Maryland, but it was already negotiating with a rival contractor. AURA then turned to Johns Hopkins, which, although it has a relatively small astronomy department, was eager to host the institute and offered the consortium low-interest state bonds for construction and a professorship for the institute's director. Additionally, it offered an isolated, wooded glade within walking distance of campus for a building site. In Janu-

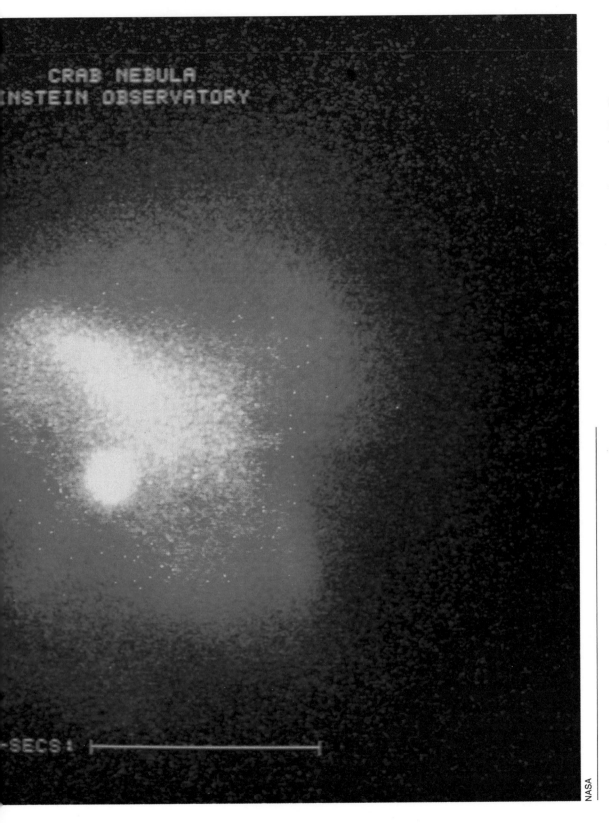

CRAB NEBULA
INSTEIN OBSERVATORY

-SECS

NASA

Because it is an entirely space-based field, X-ray astronomy has always required the sort of expensive, large-scale projects to which optical astronomers are only beginning to grow accustomed. This X-ray photograph (left) of the Crab nebula reveals the remnant of a supernova that exploded in A.D. 1054.

ary 1981 AURA got the nod from NASA and chose the university.

That June, an astronomer with extremely impressive scientific and administrative credentials, Richard Giacconi of the Harvard-Smithsonian Center for Astrophysics, was named director of Space Telescope Science Institute. Born and educated in Italy, Giacconi came to the United States in the mid-fifties and soon became involved in the emerging field of X-ray astronomy. Because X rays are blocked by the atmosphere, this is an entirely space-based science. In the 1940s and 1950s, scientists at the Naval Research Laboratories had used V-2s and other early rockets to make X-ray observations of the Sun, but their findings had sparked little enthusiasm for the field in the general astronomical community. In 1962, however, when a Giacconi-led team using a rocket that maintained an altitude of 50 miles (80 kilometers) or more for a little less than six minutes discovered the first X-ray source outside the Solar System, X-ray astronomy came into its own.

Giacconi would go on to play a leading role in the development of *Uhuru,* the X-ray satellite launched in 1970 into an equatorial orbit from a platform off the coast of Kenya. It was responsible for the first all-sky catalog of X-ray sources. Uhuru is Swahili for "freedom," and Giacconi chose the African name deliberately to defy the NASA officials who had given the mission the bland, bureaucratic title of *Small Astronomy Satellite,* or *SAS-1.*

Giacconi had also been principal investigator on the *Einstein* X-ray observatory launched in 1978 (the next year was the centennial of Albert Einstein's birth), which was used by more than 600 guest observers in the course of its two-and-a-half- year life.

Some of Giacconi's colleagues, however, find certain aspects of his background troubling. This founding father of X-ray exploration is making his optical astronomy debut with the space telescope. And he has also earned a reputation for aggressive, driven leadership. He originally wanted to name the *Einstein* observatory *Pequod,* after the ship whose final voyage Herman Melville chronicled in *Moby Dick.* For Giacconi, the name conjures up images of American Indians, New England whalers, and a spirit of

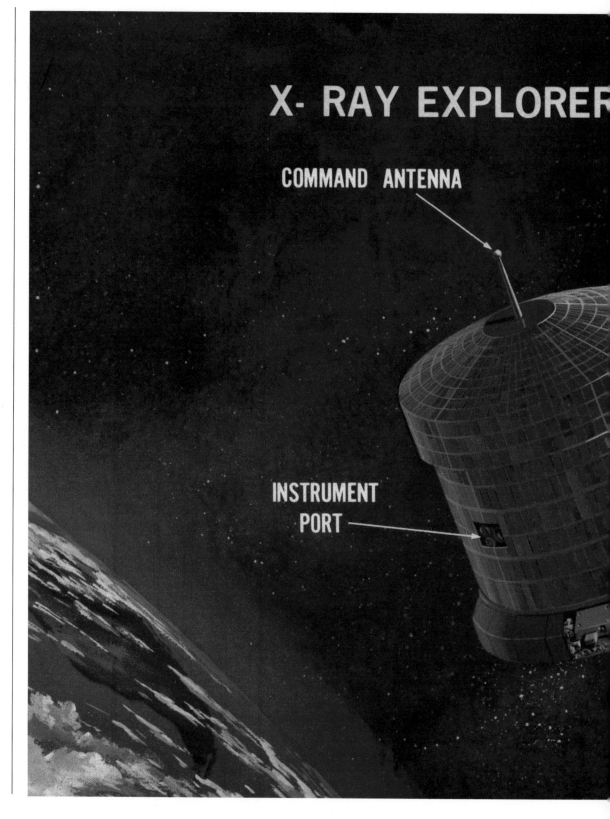

SOLAR PANELS

INSTRUMENT PORT

YO-YO DESPIN SYSTEM

NASA

Courtesy Space Telescope Science Institute

Although Richard Giacconi (above) has had a very distinguished career as an X-ray astronomer, playing a major role in both the *Uhuru* X-ray satellite (left) and the *Einstein* X-ray observatory, his appointment to head the Space Telescope Science Institute was controversial because of his lack of experience in optical astronomy.

adventure. To several of his colleagues, it recalls a doomed expedition captained by an ambitious megalomaniac.

Under Giacconi's stewardship, the science institute has grown ambitiously. Originally envisioned by NASA as employing one hundred people, the institute by 1985 had grown into an organization with approximately two hundred employees (about one-third of them scientists). The annual budget is projected to have more than quadrupled by the end of the decade, and some believe that Giacconi's ultimate goal is a staff of four hundred. A report in April 1985 by a National Research Council task force, while commending the institute's achievements, urged that future expenditures remain "within available resources."

Giacconi has made no secret of his ambitions for the institute. "What *is* this place? There's a philosophical difference," he declares. "I think NASA basically thought of the institute as a data distribution center and a service institution. But *I* conceive of it also as a first-rank research institute. I'd like to see this place like a sieve, with people coming and going, with ideas flowing. Space telescope is a terribly important resource. It goes beyond national boundaries. The limit is not observing time, but brains."

However, if observing time is not the ultimate limit to what the space telescope can accomplish, it is nonetheless an issue on the minds of a lot of the people involved with the telescope as well as those who want to be involved with it. The problem: not much time and a great deal of people who are in need of it. And among the principal tasks of the institute are maximizing that time and choosing who gets to use it.

One of the reasons why observing time is extremely limited is that the space telescope, in spite of its name, is not going very far out into space. The space shuttle will be able to place it in an orbit of about 300 miles (482 kilometers) from the Earth's surface. That's far enough to avoid almost all the distorting and filtering effects of the atmosphere, but not far enough to prevent the Earth, Sun, and Moon from coming between the telescope and its intended targets. There is nothing anyone at the institute can do about the fact that the telescope must always be pointed at least 50° away from the Sun, 15° away from the sunlit Moon, and 5°-15° away from the sunlit Earth. In the 1960s, there had been discussion about avoiding this problem by putting it into high Earth orbit or even on the Moon, but both of these alternatives are difficult, impractical, and expensive.

As head of the science institute's operations and maintenance branch, Ethan Schrier works to devise ways to enable astronomers to get as much observing time as possible out of the telescope. Schrier comes to the space-telescope project from high-energy physics, a field in which complex, expensive projects staffed by large teams are the norm. His introduction to astronomy was Giacconi's *Uhuru* X-ray mission, and he went on to work on the *Einstein* observatory.

NASA

The space telescope (left) will orbit the Earth at an altitude of approximately 300 miles (482 kilometers)—high enough to eliminate almost all of the distorting effects of the atmosphere but not high enough to prevent the Earth, Sun, and Moon from coming between the telescope and its targets. Early suggestions to place the telescope in a higher orbit or on the Moon were rejected as expensive and impractical.

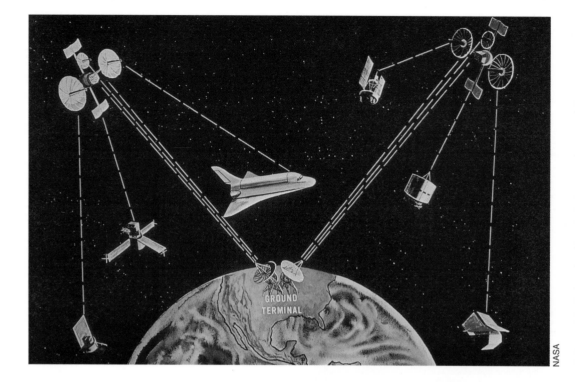

NASA

Because it is only one of several, satellites using NASA's TDRSS satellites to transmit data to the ground, the space telescope will not be in constant communication with Earth-based observers (above).

"It's not a coincidence that I came out of X-ray astronomy," he says. "My deputy also did. The optical astronomy did not have any people with experience, or at least much experience, in satellite operations. And some of the chief computer software people here came out of radio astronomy because not that many optical astronomers have experience with large computer systems."

The need to repoint the telescope following each observation places another limitation on its efficiency. The staff of the institute is working to surmount this limitation, however. "We will certainly try to optimize the science efficiency," says Schrier. "Once you build a . . . spacecraft, it certainly pays to use it as efficiently as possible, but it's going to be a very intense job coming up with an optimal schedule. There

NASA

For perhaps as long as one-fifth of each 100-minute orbit, a TDRSS satellite (left) will enable astronomers to make "real time" observations on the space telescope.

are going to be thousands of observations, so if you can make each one just five percent more efficient, the return is immense."

Most of the observations on the space telescope are preplanned, and this is not just so that Schrier and his group can arrange a schedule that permits it to move quickly from one target to the next. Because it is only one of several users of NASA's TDRSS satellites to transmit its data back to the ground, the telescope is not in continuous communication with the Earth. Most of the time, the telescope's observations are recorded on tape for later transmission to TDRSS and from TDRSS to Earth. But the observations are not entirely preplanned either. For at least some—perhaps as much as one-fifth—of each orbit, astronomers will be in contact with the satellite, able to see what the telescope sees as it sees it, able to observe in what scientists call "real time." One of the principal discoveries of astronomy in this century is how much the seemingly peaceful, orderly heavens are actually in a constant state of violent change, and real-time observations give astronomers a chance to look at the sudden eruptions of active but faint objects.

But who gets to make these scarce, precious observations at the science institute? The astronomers who developed the scientific instruments have first priority. Following launch and some initial testing, they have the telescope to themselves for the first two months and a hefty although gradually diminishing amount of the time, for a cumulative total of approximately thirty percent over the mission's first two-and-a-half years. Of the remaining seventy percent, about ten percent is director Giacconi's discretionary time, some of which is available for observations by amateur astronomers. The final sixty percent will be the prize thousands of astronomers around the world are vying for. A preliminary survey by the institute suggests that there will be about fifteen requests for time for every one that can be granted.

Neta Bahcall will play a key role in implementing the process by which that remaining sixty percent of observing time is distributed. Born and educated in Israel and married to Princeton astronomer John Bahcall (who with Spitzer has been one of the space telescope's most effective lobbyists), she is the

Although professional astronomers from around the world will be competing for observing time on the space telescope (right), Space Telescope Science Institute's Giacconi has reserved part of the time for use by amateurs.

NASA

NASA

The Space Telescope Science Institute—like its predecessor, the observatory at Goddard for the *International Ultraviolet Explorer (IUE)* (left)—will be open to scientists from all over the world.

director of the science institute's general observer branch.

Each observation is scheduled to last about half of the space telescope's one-hundred-minute orbit. And, because of time taken up by testing and the guaranteed-time observers, Bahcall believes the general observer branch will only have about 150 observations to mete out in the telescope's first year, a figure that should climb to about 250 to 300 annually over time. How will the institute deal with the problem of a limited amount of observing time desired by a much less limited group of observers?

"It involves somewhat more collaboration than typically has been done in optical astronomy, where usually one person could do a whole project," Bahcall explains. "Here, there is so much data and so much analysis and so on that you need more than one person in most cases, although there will be many small one-person projects. I expect that time on the telescope will be divided roughly into small, medium, and large projects simply to enable a large spectrum of scientific projects to be carried out."

Still, the largeness of the space-telescope project, and the largeness of the bureaucracy that has arisen to manage it—even if it is a bureaucracy of scientists—represents a major, potentially disorienting chance for optical astronomers. "Optical astronomers were the people who went up on top of a mountain and got their photographic plates and went back to their offices to sit with their pictures in their desk drawers for years and years," comments Schrier, a veteran of high-energy physics and X-ray astronomy. "Individual projects have their place and are needed," Schrier admits. "Thinking gets done individually, not in teams. But building a billion-dollar instrument doesn't get done by a couple of people calling each other over the phone and saying, 'Let's build a telescope and launch it.' It gets done by a very large group of people dealing with a government bureaucracy."

Some astronomers do not think that the telescope changes life with the astronomical community very much. "I think space telescope [is] viewed as an international observatory open to any qualified scientist who submits an excellent scientific

Technicians prepare the *IUE* before its 1978 launch. The *IUE* began the process of incorporating space astronomy into optical astronomy, one that many astronomers expect the space telescope to accelerate.

proposal," says Goddard instruments scientist Leckrone. "The *International Ultraviolet Explorer* was run that way very successfully, and that's why we have had three hundred or four hundred people a year using it. There's the same precedent in ground-based astronomy, where major, national observatories are open to any qualified person."

Still, Leckrone admits that group work may come as a new thing to many optical astronomers. "We expect space telescope to be heavily oversubscribed because there are just not enough hours in the year to give," he notes. "One way to get around that is to encourage people to get together as teams rather than as individuals competing to do the same program. That is something of a novelty. That's something that's not commonly done in ground astronomy. That's an experiment too. It'll be very in-teresting to see how effective that approach is."

Many astronomers seem almost as intrigued by the question of whether the space telescope will change the way they do their science as by the science it will enable them to do. "It certainly will change things and the changes will be significant," remarks Goddard project scientist Bogges. "Whether the changes will be all that sweeping, however, is something that I'm a little uncertain of. Before the *IUE* came along, a fair amount of space astronomy had been done; and, from a scientific point of view, the results from those satellites made fundamental changes in what we knew and believed about stars and the interstellar medium. But, from a sociological point of view, as far as astronomers were concerned, there was only a small cadre of specialists involved in space astronomy.

"*IUE* was designed as a facility that anyone could use—we took particular pains to set it up that way," explains Goddard project scientist Bogges. "You didn't have to be a specialist in space science or building satellites to take advantage of it. To a large degree, I think that's the reason space astronomy is now much more intimately incorporated into other parts of astronomy. And I think space telescope will continue that trend. There's a great deal of anticipation and curiosity about space telescope. Many astronomers will want to use it. What is not yet certain is whether we have been successful in making it so much better than what astronomers can get from ground observatories that they will want to continue using it. My hope and belief is that we have and that a few years from now you will not be able to pretend to do astronomy without using the space telescope."

NASA

O BRAVE NEW WORLDS

In December 1962, following an American failure several months earlier and a Soviet failure the previous year, a U.S. space craft, *Mariner 2,* passed within 21,748 miles (35,000 kilometers) of Venus. While the news it sent back was far from epochal, that Venus was hot and had no magnetic field, it was nonetheless information that could not have been obtained by staying on Earth. Subsequent interplanetary missions, the later *Mariners, Vikings, Pioneers,* and *Voyagers,* vastly expanded our knowledge of our neighbors in the Solar System. What can a mission that will venture no further than 300 miles (482 kilometers) from the surface of Earth have to tell us about those distance worlds it can take years to reach?

When he arrived in the early 1960s to teach and do research at the California Institute of Technol-ogy, James Westphal was a professional geophysicist with several years' experience in oil exploration as well as an amateur astronomer. At Cal Tech he found "a lot of freedom to move from one interesting thing to another" and turned his attention skyward on a full-time basis. Westphal took the skills and the technology he had used to study Earth and applied them to objects in The Solar System, to Jupiter and the Moon in particular. With a colleague from the nearby Jet Propulsion Laboratory, he pioneered the use of the deep infrared (wavelengths of about 2 microns, just slightly larger than visible light). Among his discoveries was that the Moon's surface is solid rock and not, as some in those pre-*Apollo*-landing days feared, a kilometer or two of fine dust.

Westphal is the head of the Cal Tech team that designed and devel-

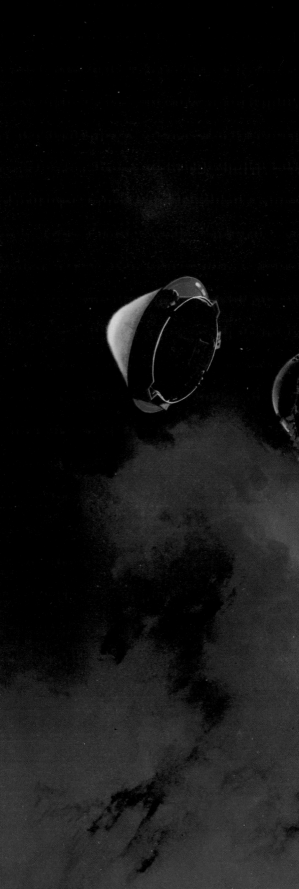

NASA

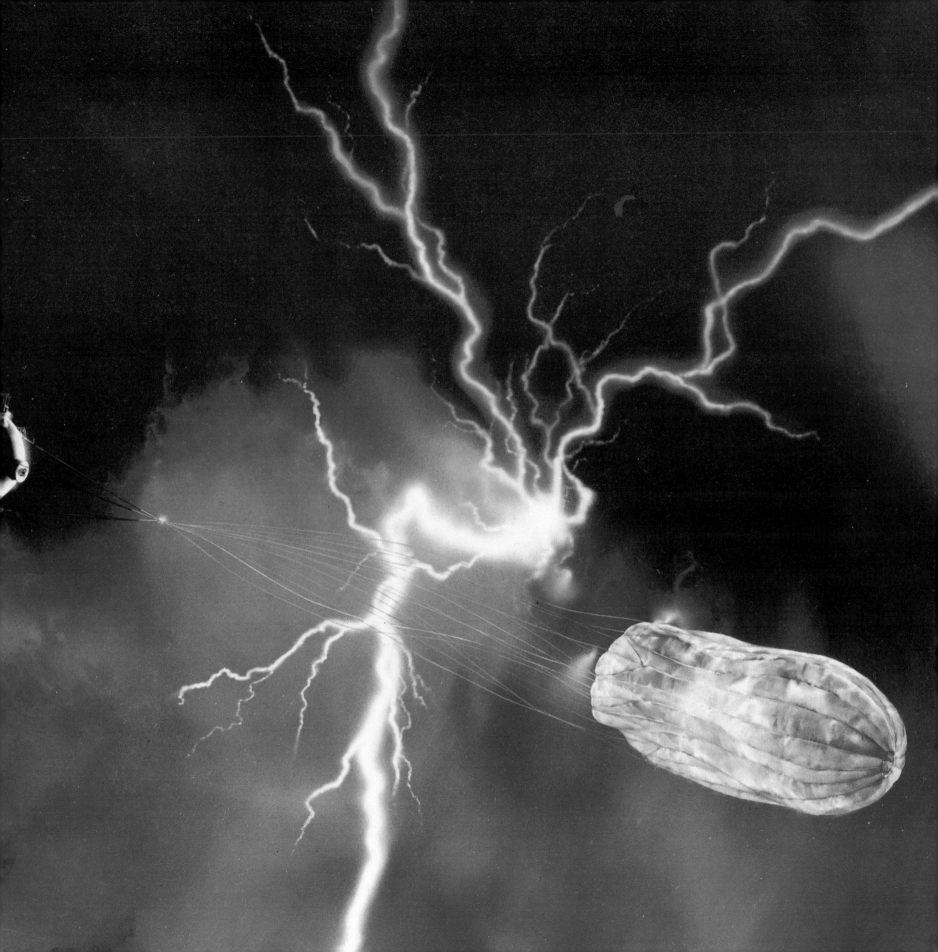

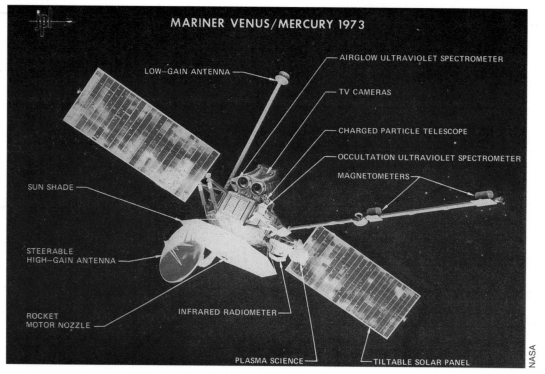

MARINER VENUS/MERCURY 1973

LOW–GAIN ANTENNA

AIRGLOW ULTRAVIOLET SPECTROMETER

TV CAMERAS

CHARGED PARTICLE TELESCOPE

OCCULTATION ULTRAVIOLET SPECTROMETER

MAGNETOMETERS

SUN SHADE

STEERABLE
HIGH–GAIN ANTENNA

ROCKET
MOTOR NOZZLE

INFRARED RADIOMETER

PLASMA SCIENCE

TILTABLE SOLAR PANEL

NASA

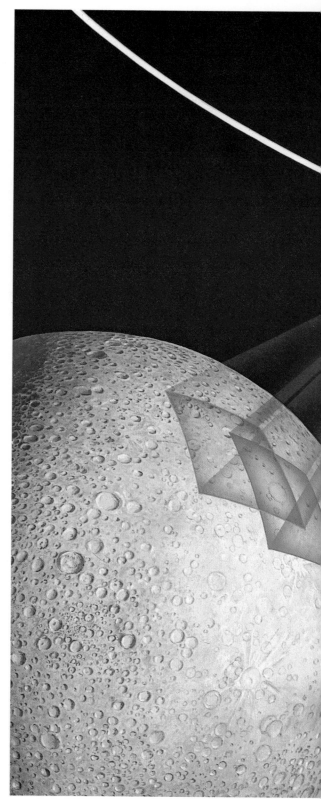

Flyby spacecraft have greatly added to our knowledge of the planets. The *Mariner 10* mission (above) produced some of the first good photographs of the desolate landscape of Mercury. A future flyby, *Galileo* (shown in an artist's conception on the previous page), should add greatly to our knowledge of Jupiter's atmosphere.

Mariner 10 (shown at right approaching Mercury in an artist's conception), however, did not succeed in photographing the dark side of the closest planet to the Sun. The space telescope may attempt that tricky maneuver after it has been in orbit for a few years.

NASA

oped the wide-field planetary camera (with its then highly innovative use of CCDs) for the space telescope. What will the camera in its planetary mode see that the flyby missions missed? "One advantage is that we are going to have long periods of time in which to watch things change," Westphal says. "And the fact that we can look deep into the ultraviolet with good resolution is also a big improvement."

Mercury, the innermost and fourth-brightest planet (almost as bright as that brightest of stars, Sirius), is for all its brilliance singularly difficult to observe clearly. The problem is the company it keeps: the Sun. Mercury appears near the horizon (which tends to distort observations) at dawn or dusk (which keeps observations brief). Earth-based optical observations of this airless, cratered world (much like our Moon) are so difficult that the length of the Mercury's rotation on its axis—58.6 days, or two-thirds the time it takes to orbit the Sun (scientists now assume that this ratio is not an accident although the reason for it is uncertain)—was not tracked accurately until established by radar observations in 1965.

And today, even after the splendid photographs sent back to earth by *Mariner 10* in 1974, much about the planet remains unknown. In fact, half of Mercury has yet to be observed. Westphal hopes that the space telescope will rectify this situation. He explains, "It's a tricky thing. Because it's so close to the Sun, the only way we'll be able to take a picture of it is, when the satellite is still in the Earth's shadow,

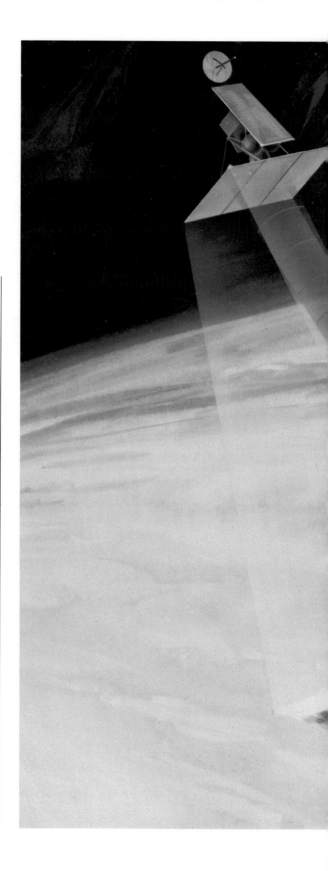

An artist's conception (right), illustrates how the planet's thick cloud cover may force scientists to use such means as radar to explore the Venusian surface.

to point the telescope at the position in the sky where Mercury is going to be when the satellite comes out from the shadow. Then, as Mercury comes over the horizon, take some pictures, and quickly point the telescope in another direction before the Sun comes up. That's pretty sporty. We won't try it until after we have some real experience with the spacecraft, perhaps two or three or four years into the mission." The results of this process will not produce *Mariner*-quality images, but they will give astronomers a chance to see if the entire planet is as lunar as the flyby pictures suggest. Westphal expects that all of Mercury will be of a consistent lunar desolation, but he is willing to be surprised. "When we flew *Mariner 4* by one side of Mars all we saw were craters," he recalled. "Later, with *Mariner 6* and *7,* we saw the other side, and there were volcanoes, 'Grand Canyons,' and all kinds of stuff."

The space telescope will tell astronomers nothing about the surface of Venus. Although Venus is Earth's closest neighbor and, after the Sun and Moon, the brightest object in the sky, the second planet's surface is completely obscured by a heavy cloud cover made up of sulfuric acid droplets. No surface markings or distinct clouds can be seen in the optical range. No details could be photographed until astronomers attempted to observe Venus in the ultraviolet, which came as a considerable surprise, since ultraviolet is completely useless in photographing clouds on Earth. What astronomers then saw were great cloudlike formations changing daily and moving in a four-day rotation. Later radar observations determined that this cloud movement did not reflect the planet's rotation, which takes 243 days and occurs in the opposite direction of all other planetary rotations. These ultraviolet clouds, whose movements result from complicated circulation patterns in the Venusian atmosphere, are what the space telescope will look at. "We'll be able to take pictures that are going to look very much like the ones that were taken by *Mariner,*" says Westphal, "but we have the huge advantage that we'll be able to

NASA

This ultraviolet view of Venus (above) was taken from 460,000 miles (742,000 kilometers) by *Mariner 10*'s television cameras one day after the spacecraft flew past the planet on its way to Mercury. The clouds change daily and move in a four-day rotation.

NASA

take them for years and years."

The surface features of Mars, by contrast, are visible to any earthbound observer with a trained eye and small telescope. Because Mars is relatively close to Earth without being too close to the Sun or blanketed by an obscuring atmosphere, it has been well observed by astronomers. Scientists have, in fact, been tempted to look too hard at this fourth planet. In the early years of this century, the great American astronomer Percival Lowell claimed to have seen a network of canals, obviously the handiwork of some intelligent life form, girding the planet. Lowell was wrong; the canals were the projection of an orderly imagination on a vaguer reality. On the other hand, Lowell was absolutely correct in another of his predictions. In 1914, he foretold the existence and orbit of Pluto; it took, however, another sixteen years to develop the astronomical technology that would actually locate this most distant planet.

For Mars, again, the space telescope's major advantage will be the length of time over which it will be observing. This should enable it to compile a detailed record of two related and highly distinctive features of the Martian climate: the planet's polar caps and its massive, planetwide dust storms. It will not, however, produce images to rival those from the missions that actually traveled to the red planet. "We won't be able to get resolution anything like what the *Viking* spacecraft did," admits Westphal.

He has considerably higher hopes for Jupiter, the fifth planet

NASA

An artist's rendering of a Martian volcano (above) based on a photograph taken by *Mariner 5* in 1972 shows a close-up view of Caldera Nix Olympica. So far the largest known volcano in the Solar System, it measures 375 miles (604.8 kilometers) across at its base.

A dramatic photograph of Mars taken by a *Viking* spacecraft is pictured here (right). The starry background is the work of a NASA artist.

NASA

This image of Saturn (right) is a color-enhanced photograph taken by *Voyager 1* in 1980. The still mysterious ''spokes'' are faintly visible in the lower right portion of the rings. A color-enhanced photograph of the underside of Saturn's rings (below) was also taken by *Voyager 1* in 1980.

from the sun. ''In the case of Jupiter,'' Westphal says, ''the pictures are going to be quite spectacular, quite like the pictures *Voyager* took about five days from its encounter with the planet.'' The pictures should help astronomers and meteorologists clear up some of the mysteries of the Jovian atmosphere. Like Venus, the surface of Jupiter is completely shielded by a thick layer of clouds. But unlike the colorless Venusian cloud bank, the giant planet's cover is multihued, its most striking feature being a giant red spot, which remains almost as great a puzzle to scientists today as it was when first discovered in

NASA

NASA

Jupiter (left), with its prominent red spot, was photographed from an Earth-based telescope in 1966. A composite of the largest planet (below) made from photographic images taken by *Voyager 2* exaggerates the color differences in the Jovian atmosphere to make details of the cloud structure more apparent.

1830 (which makes it a long-lived atmospheric condition).

Says Westphal, "The little samples that we have from *Voyager* are fascinating, but what does Jupiter look like in between? We've seen that it changes radically from one spacecraft visit to another, but how does that change happen? We can't really tell from the ground.

"In all of these cases—Venus, Mars, and Jupiter—the fact that we can work in the ultraviolet with special resolution is a major improvement. What will we see there? I don't know. No one has ever looked at Jupiter in the ultraviolet with enough resolution to see much of anything. If you assume that the atmosphere is well behaved, then we won't see anything. But, personally, I'd be absolutely amazed if we didn't see some interesting things."

Although Saturn, along with the other planets beyond Jupiter (Uranus, Neptune, and Pluto) will also be looked at with the planetary camera, the resolution will be not nearly as good as that obtainable from a flyby.

Nonetheless, data from the space telescope may help explain one of the most alluring riddles of the Solar System: the rings of Saturn. While rings have recently been discovered around Jupiter and Uranus

Courtesy British Aerospace

as well, Saturn's are by far the most brilliant. And, like Jupiter's atmosphere, it seems the more we learn about these rings, the more complex, unpredictable, and just plain strange they seem to be. Consider the "spokes" in the rings. *Voyager 1* discovered and *Voyager 2* confirmed the existence of rotating spokes—which look like large, dark fingers—in the rings. "The one thing that we can do quite well with Saturn is to monitor this spoke business," says Westphal. "I think everybody would appreciate a really nice set of sharp pictures over a long period."

Planets and their moons are not the only occupants of the Solar System. Asteroids, that belt of tiny rocklike solar satellites between Mars and Jupiter, are another class of permanent residents. And there are also comets—flashy, fiery balls of ice and dirt with elongated orbits that take them both nearby and far away from the Sun. Launch delays prevented the space telescope from observing Halley's comet, the most famous of these phenomena. What was lost was not so much the close-

up views, the sort of images that the European *Giotto* mission took so spectacularly, as the chance to view the comet when it was just beginning its approach.

The planetary camera is not the only instrument that will be used to observe the planets. The high-resolution spectrograph would have been the key instrument in viewing Halley's comet had the telescope been in orbit at the time. It will probably play at least as important a role in observing planetary atmospheres as the camera. And the Bless-designed high-speed photometer will also make several important planetary observations.

The photometer, a light-measuring device capable of registering up to 100,000 variations a second, will use the variations of light that result from occultations (that is, when one object passes in front of another) to make its observations. It will measure, for example, the changes in light that occur when a star passes behind Saturn's rings.

"With the planetary camera," Bless says, "you can get resolutions of maybe 0.1 of an arc second,

Europe's first deep space explorer, the Giotto Interceptor, designed and assembled by British Aerospace, was launched on July 2, 1985, in time to make its historic observations of Halley's Comet in March, 1986.

which is about ten times better than you can get from a ground-based instrument. And on Saturn 0.1 of an arc second corresponds to about 600 kilometers [373 miles]. With occultations, watching the rings go in front of a star with the high-speed photometer, we can get a resolution of about one kilometer [0.62 miles]."

The disadvantage of the photometer, Bless notes, is that it gives one-dimensional readings, not two-dimensional pictures like the planetary camera. "The camera has the huge advantage in that it can take a photograph of the whole system, while we just have to take potluck. On the other hand, we expect to get an occultation per year, and, over a few years, we will be able to look at the dynamics of the rings in very great detail."

The photometer will also use occultations to examine the upper atmosphere of Jupiter. "When Jupiter moves in front of a star, the star's light will fade slowly because it's passing through a thick atmosphere," explains Bless. "This enables you to get the structure, the actual temperature and density of the upper atmosphere. You can also get the helium abundance. And, unlike with a flyby such as *Voyager*, when you take pictures for a few days and that's it, we'll be able to watch how things like Saturn's rings and Jupiter's atmosphere evolve over time."

There also exists the possibility that the space telescope may find a new world to explore. Not, in all likelihood, in our Solar System, which has been pretty thoroughly searched by astronomers in the

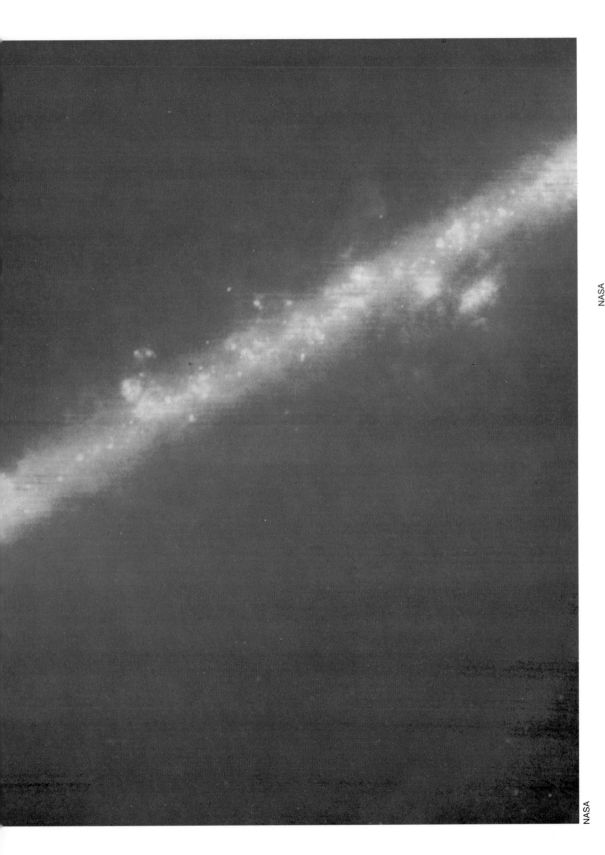

NASA

NASA

Acolor-enhanced composite photograph of Saturn (above) taken by *Voyager 2* combines three images taken in 1981. The space telescope, in addition to observing the planets in our Solar System, will also be on the lookout for as yet unknown planets orbiting around other stars. A photograph of the Milky Way Galaxy's center (left) taken by the *Infrared Astronomical Satellite (IRAS)* offers a clearer view than do optical telescopes.

NASA

half century since Pluto's discovery. No, the space telescope will be looking for a planet in orbit around some other star than our Sun.

The search is not a new one. In 1937, American astronomer Peter van de Kamp began a study of Barnard's Star, the closest single star to the Sun. (The closest star, Proxima Centauri, is part of a three-part star system along with the next-closest star, Alpha Centauri, which is in fact made up of two stars, Alpha Centauri A and B.) His goal was to observe the star with greater accuracy than had ever before been attempted to see if there was any wobble in its flight path as it moved across the sky over several decades. Wobble he looked for, and wobble he found. In 1963, van de Kamp announced that based on his observations he concluded that a Jupiter-sized planet was circling Barnard's Star in a twenty-four-year orbit. Six years later he revised his findings. He then concluded that Barnard's Star must have two Jupiter-sized companions. Unfortunately, van de Kamp, like Lowell before him, appears to be guilty of reading too much into his observations. In the

1970s, studies showed that a good deal of the wobble he had seen was the result of either alterations made on his telescope, the 24-inch (60-centimeter) refractor at the Sproul Observatory outside Philadelphia, or the telescope's basic inadequacy. There may be planets orbiting Barnard's Star, but no one on Earth has any evidence of their existence.

This is the reason that F. Ducchio Macchetto is so interested in finding such a planet. Although Macchetto was born in the northern Italian city of Turin, his family joined the post–World War II migration of Italians to Argentina when he was still a young child. Macchetto began studying astronomy in Argentina at the University of Cordoba, a school with a glorious past. It was there in the late-nineteenth century that the first systematic survey of the Southern Hemisphere's sky began. But, during Macchetto's student days, the school was experiencing problems. "Our professors would come from Buenos Aires, about 700 miles [1,126 kilometers] away, to teach us over the weekend," he recalls. "We were all very dedicated—professors

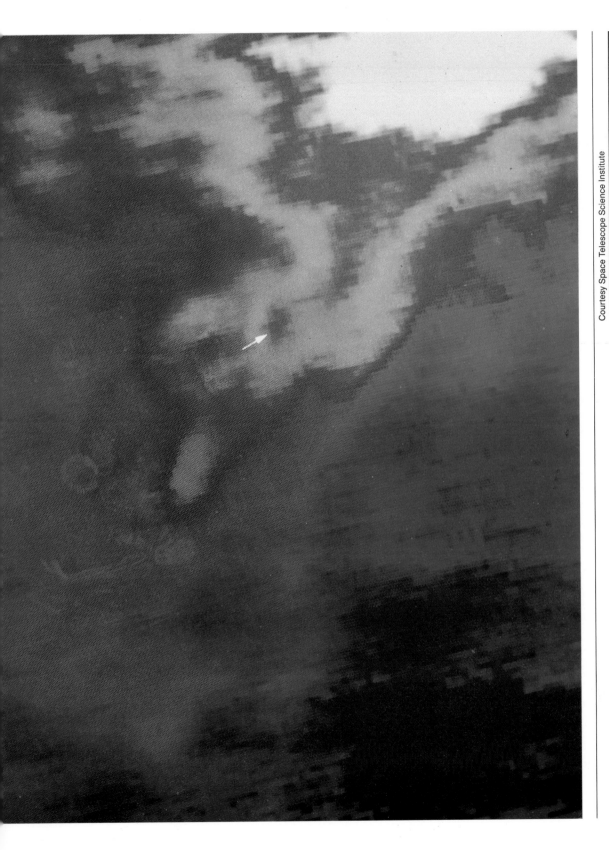

Courtesy Space Telescope Science Institute

F. Ducchio Macchetto (above) is the European Space Agency's representative to the space telescope project and is the project scientist for the faint-object camera.

A recently discovered newborn star (the dark patch pointed out by the arrow) is seen amongst a cloud of dust and gas called Barnard 5 in this image (left) produced by data from the *IRAS*. The infrared satellite also saw dust clouds around some nearby stars, which astronomers think, on closer examination, could turn out to be planetary systems.

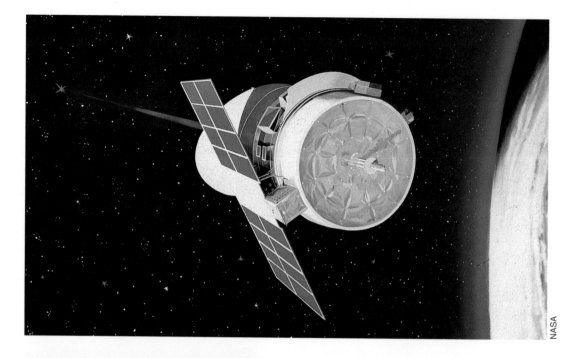

An artist's conception of the *IRAS* (right) shows the satellite in Earth orbit. An artist's view of two *IRAS* findings (below) reveals how it is possible for a collection of particles to share a common orbit around the sun and how the collision of a comet and an asteroid could create a cloud of debris.

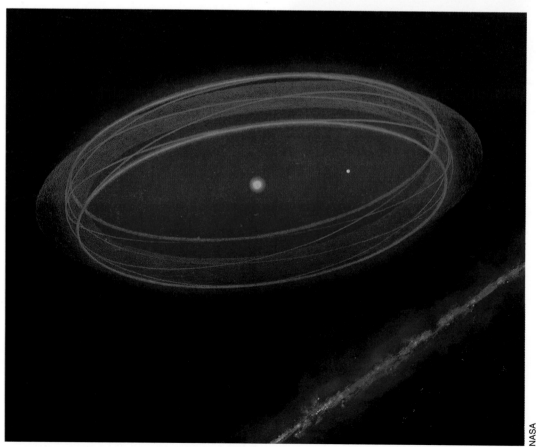

A Delta launch vehicle (right) takes off from Vandenburg Air Force Base in January 1983 to carry the *IRAS* into orbit.

and students—but the professors were telling us that as soon as you get your degree go away because physicists and astronomers have very little future in this country. About two-thirds of those who graduated with me left. Some of them returned later, but most of us left and never returned." Macchetto went to Rome where he completed his doctoral studies in astronomy and resettled in his native Italy.

He has several roles in the space-telescope project. He is director of the science institute's instrument-support branch, in which capacity

NASA

he supervised the prelaunch calibration of the instruments. He is the ESA's representative to the space-telescope project. And he is the developer of and principal investigator on the instrument the Europeans built for the telescope, the faint-object camera. It is in this final role that he hopes, like an earlier Italian, Christopher Columbus, the Admiral of the Ocean Seas, to find new worlds.

"Observations made by *IRAS* [the *Infrared Astronomy Satellite* built by the United States, the United Kingdom, and the Netherlands that flew in 1983] have identified rings or dust features around nearby stars," Macchetto explains. "So we know that around many stars close enough for inspection there are at least the makings of planetary systems. The inference is that some, if not all, of these stars could have a planet circling. I'm not overly optimistic that we can find twenty, but if there are one or two out there, and they are the size of Jupiter and are about the same distance from the Sun that Jupiter is, and they are traveling around a star that is within 10 light years away from us, then we have a good chance of being about to see them. The wide-field camera will be used to look for star wobble, but we are better suited to look at the planet directly. There is a special feature on the faint-object camera—we call it a coronagraphic finger—that can block out the light from the star. We can then increase the contrast on the planet because we're blocking out the light from the star. We'll only see the planet as a speck of light which we won't be able to see features on. But, scientifically it will be a great thing."

THE UNIVERSE ACCORDING TO HUBBLE

The astronomers who have had space missions named after them can be counted on the fingers of one hand. As mentioned earlier, the fourth of the Orbiting Astronomical Observatories was named *Copernicus* to mark the five-hundredth anniversary of the Polish monk who dared to suggest that the Earth—and the other five known planets—traveled about the Sun. The X-ray observatory launched in 1978, the year before the centennial of the birth of Albert Einstein, was named for that great physicist, who, while not himself an observational astronomer, offered theories that vastly expanded our knowledge of the heavens, from the orbit of Mercury to the shape of the universe. A forthcoming planetary probe to the outer planets is named *Galileo*, after the first scientist to direct a telescope at the night sky. And, to this exclusive list, the addition of

the name Edward Powell Hubble, commemorated by the Hubble Space Telescope, is entirely appropriate. The space telescope became the Hubble Space Telescope in the mid-seventies, but was not named to mark a centennial.

Born in Missouri in 1889, Hubble attended school in Kentucky and Chicago, before receiving a scholarship to the University of Chicago, where he was both an exceptional student under such eminent teachers as physicist Robert Millikan and astronomer George Ellery Hale and a promising athlete. After he graduated, a boxing promoter offered to train him to face world-champion heavyweight Jack Johnson (the legendary black fighter whose career is chronicled in the play and movie *The Great White Hope*), but Hubble decided to go to Oxford University on a Rhodes scholarship instead.

At Oxford, Hubble kept up with

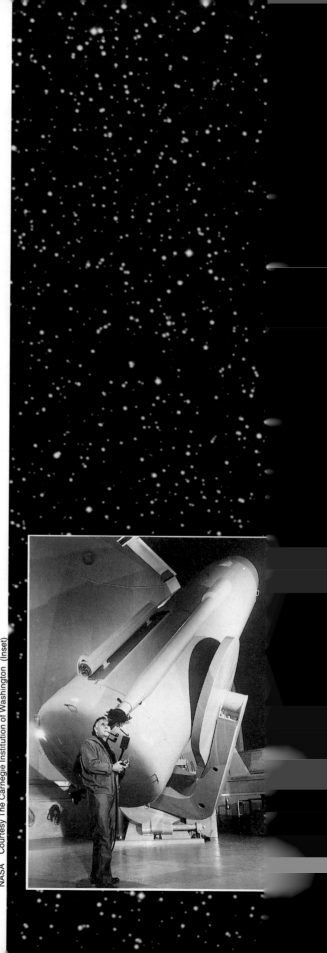

NASA   Courtesy The Carnegie Institution of Washington   (Inset)

The nebula in the constellation Andromeda (shown on the previous page in a photograph from an Earth-based telescope) is visible to the unaided—but well-trained—eye and has been known since A.D. 964. However, scientists did not know what nebulas were until Edward Powell Hubble (inset on the previous page) determined in 1924 that they are galaxies like the Milky Way.

The Ring nebula (right) photographed from an Earth-based telescope is one of the best known examples of a planetary nebula. Hubble used the new 100-inch (2.5-meter) telescope at Mount Wilson to resolve a nebula into individual stars.

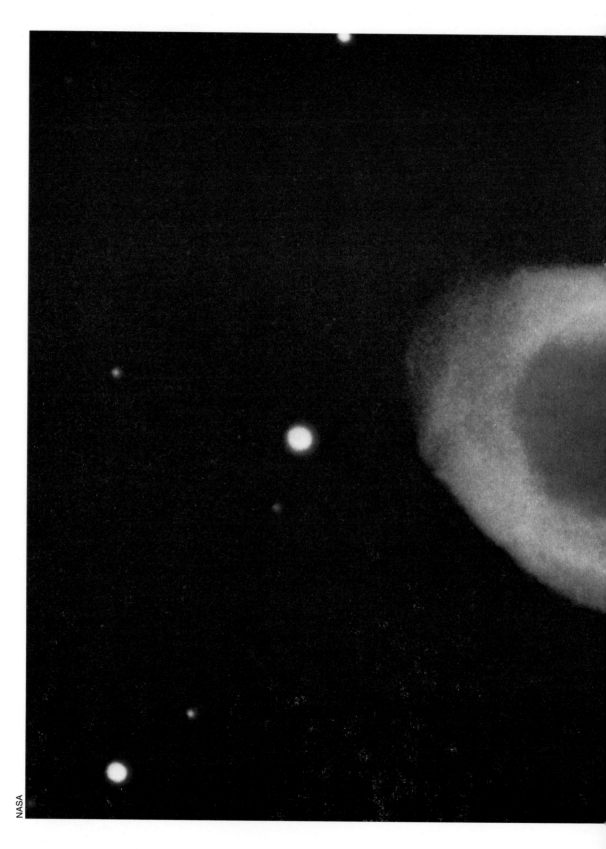

NASA

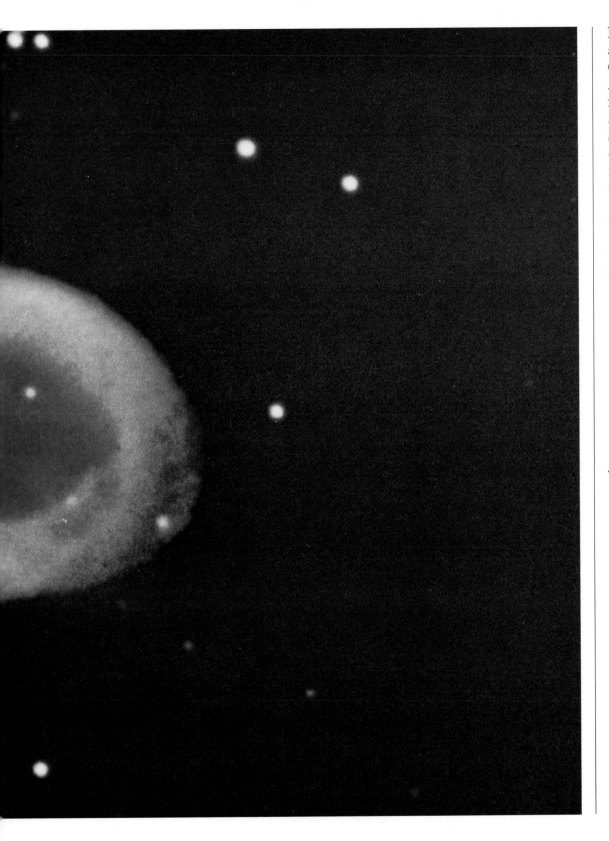

his boxing as an amateur, fighting an exhibition match with George Carpentier, the French champion. Academically, however, he switched fields from mathematics to law, and, upon his return to the United States in 1913, he set up a law office in Louisville after being admitted to the Kentucky bar. The defection from science was short lived. In 1914, Hubble was back with the University of Chicago doing graduate work in astronomy at the school's 40-inch (l-meter) refracting telescope at the Yerkes Observatory, where he attracted the attention of his former mentor Hale. In the meantime, Hale had become director of the Mount Wilson Observatory near Los Angeles. Hale offered Hubble a position at Mount Wilson. However, the United States entered World War I and Hubble entered the army (where he rose from private to major), but not before cabling Hale to say that he would accept the job as soon as he was out of the service.

When Hubble arrived at Mount Wilson in 1919, the observatory's new 100-inch (2.5-meter) reflecting telescope, the largest in the world at that time, was just coming into operational use. The timing was fortunate for the young astronomer, for he would need its full power to determine, for the first time, the true nature of those fuzzy patches of light in the sky that were evidently not stars and which astronomers called "nebula," from the Latin word for cloud.

At least one nebula, the one in the constellation Andromeda, is visible to the unaided but experienced eye on a clear night and is

Mary Evans Picture Library

Thomas Wright's observations of the Milky Way in 1750 led him to propose the configuration of stars in what would later be called a galaxy (left).

In 1828, William and Caroline Lucretia Herschel (right) collaborated to complete a catalog of 2,500 stellar clouds that they had identified.

mentioned in the *Book of Fixed Stars*, written in A.D. 964 by the Persian astronomer Al-Sufi, who described it as a "little cloud." When astronomers began searching the skies with telescopes in the seventeenth and eighteenth centuries, they started to find nebulas with increasing frequency (and annoyance, since they were much more interested in discovering comets). In 1781, comet-hunter Charles Messier published a catalog of the 103 nebulas that had been discovered up to that point, and astronomers still refer to these celestial objects by the numbers he assigned them. The Andromeda nebula, for example, is M31. Meanwhile, more and more of these stellar clouds were discovered. In 1828, Caroline

Herschel completed a catalog of the 2,500 she had identified in collaboration with her brother, William Herschel. And by the early twentieth century, astronomers suspected the existence of as many as half a million. But what were they? And what were they made of?

Among the first of Galileo's astronomical discoveries was that the Milky Way, the bright band of light across the night sky, is composed of a great many stars. In 1750, Thomas Wright, an English theologian and scientist, suggested that the appearance of the Milky Way was evidence that the stars were arranged in a flat, disklike structure. We now call such a structure a galaxy, from the Greek word *gala*, or "milk." The idea that at least

some nebulas were galaxies, "islands of stars," was put forward in the late eighteenth century by both the German philosopher Immanuel Kant and the English astronomer William Herschel, for whom nebulas were a virtual obsession. Still, as late as 1920, an acrimonious debate on the subject erupted at an annual meeting of the National Academy of Sciences, with one side arguing that some nebulas were galaxies like the Milky Way and the other maintaining that all were just "truly nebulous objects" within our galaxy.

With the new reflector at Mount Wilson, Hubble was able for the first time to resolve the Andromeda nebula into individual stars. He observed several stars in Andromeda

Mary Evans Picture Library

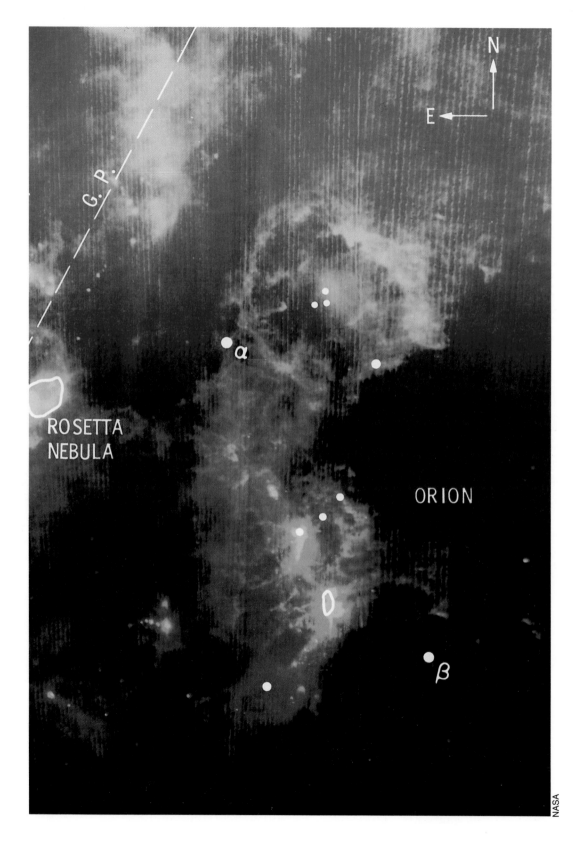

N

E

G.P.

ROSETTA
NEBULA

ORION

α

β

NASA

At right is Galaxy NGC 5907. It is an example of a spiral disk galaxy with a central bulge and arms that are studded with newly forming stars. An image of the constellation Orion and the Rosetta nebula (left) taken by the *IRAS* reveals a clearer view than that seen from optical telescopes. Note that these star formations are outside of the Milky Way (shown crossing the upper left corner).

NASA

of the type known as Cepheid variables, whose absolute brightness can be determined by how long they glow between dimouts, and whose distance can be determined by comparing their absolute brightness with their apparent brightness—that is, how much of their light reaches the Earth. Through the results of his observations, Hubble was able to show that at least some nebulas are not only composed of stars but are so distant that they must lie outside the Milky Way. When Hubble's findings were presented to a meeting of the American Astronomical Society in 1924, all present realized that the galaxy-nebula debate had been settled, and a new era in astronomy was beginning.

Hubble continued his observations of galaxies, producing a system of classification for them that remains in use. He also sought to determine their distance and the speed at which they were moving. For distances, Hubble used Cepheids. For speeds, he used the Doppler effect on the galaxies spectra. Discovered by Christian Johann Doppler, a nineteenth-century Austrian physicist and mathematician, the Doppler effect can be observed by standing next to a highway: The noise of an approaching car sounds more highly pitched than the noise from the same car as it moves away. Shortly after Doppler published his discovery, a Dutch scientist staged one of the most charming experiments in the history of science by hiring an orchestra of trumpeters to play in an open railway car speeding through the Dutch countryside. Stated more abstractly, Doppler observed that light or sound from a source moving toward an observer has a shorter wavelength than the same object at rest, and that the opposite is also true. Astronomers use the Doppler effect when they analyze light spectra: The premise is that light from approaching objects shifts toward the blue and from retreating objects toward the red.

By 1929, Hubble had examined the distance and speed of forty galaxies and come to a remarkable conclusion. With the notable exception of a few nearby galaxies, like the one in Andromeda, all galaxies were red shifted, that is, moving away from us. And the further away they were, the faster they were moving. The universe appears to be expanding. This has led many astronomers from 1929 through to the present to embrace what has become known as the big bang theory of the universe, the idea that the universe began in a tremendous explosion about fifteen billion years ago. This estimate of fifteen billion is based on the rate at which the expansion of the universe is believed to be accelerating, a rate known as Hubble's constant and currently believed to be about 9.3 miles (15 kilometers) per second per million light years.

As for Hubble himself, he continued his observations of galaxies with Mount Wilson's 100-inch (2.5-meter) reflector, and, when the 200-inch (5-meter) telescope at Mount Palomar became operational in 1949, he was the first one to use it. Hubble died in 1953.

The Hubble Space Telescope is expected to be a major new tool for astronomers in exploring the universe according to Hubble, an evolving cosmos in rapid expansion. The range of planned observations is immense. The team that developed one of the scientific instruments compiled an initial "wish" list of forty desired observations, an amount that would take considerably more than the limited amount of observation time the team was guaranteed. After months of committee meetings, the team pared its list down to thirty observations.

Goddard's Leckrone plans to investigate an aspect of cosmic creation. He wants to use the high-resolution spectroscope to determine the abundance of the chemical elements in stars of the Milky Way, which astronomers, being, after all, anthropocentric humans, refer to as the Galaxy. "There are various types of questions that you can have hopes of answering based on some data about the chemical compositions of stars and how compositions vary with the ages of stars," he says. "Ultimately, the main questions I'm interested in have to do with the origin of the chemical elements as the Galaxy itself has evolved. Where does the iron in the human body come from? Why is a particular element overa-

NASA

Galaxy NGC 7331 (above) is an example of a spiral galaxy, characterized by arms that are studded with bright, newly formed stars.

An artist's conception depicts a spherical halo of neutrinos around our galaxy (right). New evidence suggests that these subatomic particles may have been produced in the first few moments following creation.

NASA

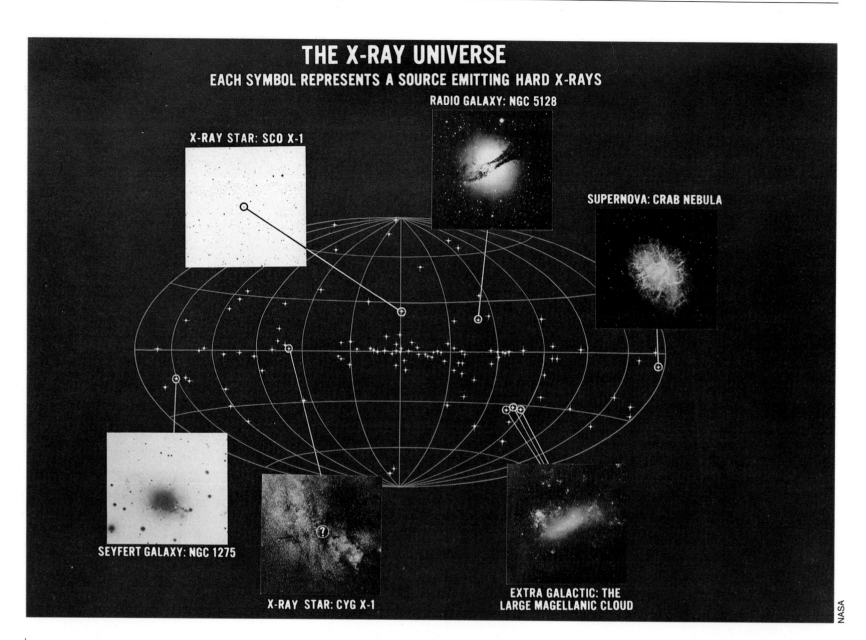

THE X-RAY UNIVERSE
EACH SYMBOL REPRESENTS A SOURCE EMITTING HARD X-RAYS

RADIO GALAXY: NGC 5128

X-RAY STAR: SCO X-1

SUPERNOVA: CRAB NEBULA

SEYFERT GALAXY: NGC 1275

X-RAY STAR: CYG X-1

EXTRA GALACTIC: THE LARGE MAGELLANIC CLOUD

NASA

bundant or underabundant in a particular type of star?"

As director of the science institute's research support branch, Peter Stockman expects the first year or so that the telescope is in orbit to be supervising performance tests on the satellite. Eventually, though, he hopes to use the space telescope to learn about star death. In several billion years, our Sun will begin its own quiet demise, turning color from orange to red, expanding to about four hundred times it present diameter to engulf all the inner planets through Mars, to become a red giant. Then, as its thermonuclear fires grow cooler, the Sun will collapse in on itself to become a white dwarf—that is, a star of the Sun's mass, Earth's diameter, and a density a million times greater than water. It will not be the first time this has happened to a relatively small star like the Sun. Astronomers estimate that there are ten billion white dwarfs in the Milky Way.

"My project's a very simple one," Stockman says. "One of the things it hasn't been possible to do from space is to look with good time resolution—that is, to see change

The data and initial observations (left) derived from the first X-ray satellite, *Explorer 42*, have grown in number and quality over the years. About 116 X-ray objects have been seen, which is about three times the number previously observed.

A false-color isophote of the Sun taken from *Skylab* (right). In several billion years, the Sun will begin its quiet demise, turning from orange to red and expanding, then collapsing in on itself to become a white dwarf.

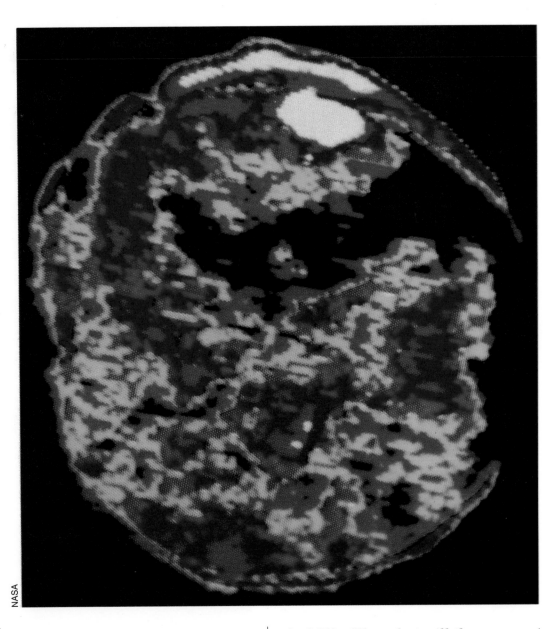

NASA

on a second-by-second or even minute-by-minute basis—at ultraviolet light from stars. There are a class of stars in the broad class of cataclysmic variables, which are binary star systems with a white dwarf being circled by a smaller star. In the systems I want to look at, there are very hot spots on the white dwarf where matter is falling onto it from the second star, and these spots rotate as the dwarf rotates. By looking at the rotation of these spots with the faint-object spectrograph, I'll learn a lot about the size and geometry of the dwarfs."

The stars that Stockman wants to look at are about 200 or 300 light years away. To astronomers, that means they are our intragalactic next-door neighbors. Outside the Milky Way, what will the space telescope look at? Other galaxies, of course. One galactic group that has particularly intrigued scientists are active galaxies, that is, those which give off particularly large amounts of energy.

"One of the most interesting fields of research in astrophysics is what happens in the nuclei of galaxies that are active sources of ra-

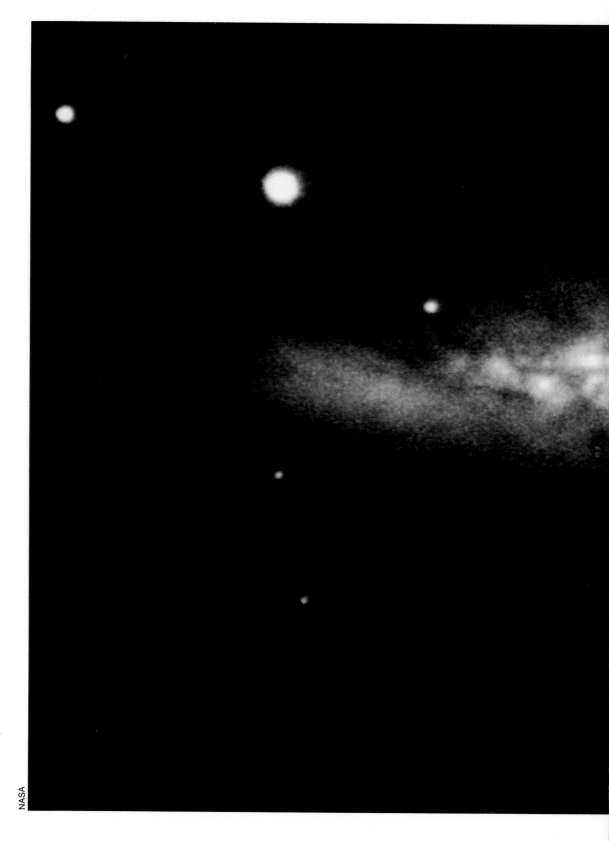

Galaxy M82 is one of a small minority of galaxies with an active nucleus. Astronomers hope to get a better view of these galaxies with the space telescope than is possible from the Earth and are particularly interested in those galaxies with long, jet-like features extending from their nucleuses.

NASA

dio energy," notes Macchetto, the man of many roles at both the ESA and the Space Telescope Science Institute. "From the center of these galaxies, we've seen jetlike features—long, thin lines of light—come out over large distances, from thousands of light years to hundreds of thousands of light years. The jets are much easier to observe well from space than from the ground. Because they are so narrow, perhaps 1 arc second in width although many arc seconds in length, you need very high resolution. And because these jets emit a great deal of ultraviolet energy, while radio galaxies tend to die down in the ultraviolet, you can get much better contrast in that wavelength. By observing from space with the faint-object camera, the contrast of the jet to the underlying galaxy is increased by two or three orders of magnitude."

Macchetto explains that astronomers believe that the function of the jets is to carry energy outward from the galaxy's central energy source. What they don't understand is the nature of that powerful source itself. "By studying the behavior of these jets and these galaxies in the optical, we hope to understand the characteristics of the central energy source," Macchetto says. "Is it a black hole? We do not yet know and need observations, particularly in the optical wavelength, to unravel this mystery."

Black holes are among the most mystifying objects in contemporary astronomy, in part because, although many scientists believe they exist, no one has yet seen one. While they might possibly turn out

Courtesy AT&T Bell Laboratories

to be a product of an overcredulous imagination, like Lowell's Martian canals, there is a long history of successful predictions in astronomy. Cases in point are Lowell's prediction of the existence of Pluto and the mid-nineteenth-century prediction that Sirius—the Dog Star, the brightest star in the night sky—has a small companion (nicknamed the Pup), eighteen years before that companion was sighted. Black holes, however, will never be "sighted." By definition, they are stars at the greatest conceivable stage of collapse, stars that have collapsed into structures at once so massive and so compact that no light or any other radiation can escape from them.

How will they ever be found? "The high-speed photometer team is going to look for the so-called signature of a black hole," explains Bless, the instrument's principal investigator. "What we mean by that is that the calculations suggest that matter falling into a black hole will orbit the black hole very quickly, and as it does it will emit radiation in bursts, and these bursts will both change in their period and their brightness in a way that occurs only when something falls into a black hole. If we're lucky, we might be able to measure these bursts of radiation as a glob of matter falls into a black hole. This would help pin down the existence of a black hole."

E. Margaret Burbidge, a professor at the University of California at San Diego and principal investigator on the space telescope's faint-object spectrograph, will also be

In the 1930s, Bell Labs engineer Karl Jansky built the rotating antennae (left) in order to find the source of radio waves that were interfering with radio telephone reception. He eventually made the important discovery that they were coming from the center of the Milky Way.

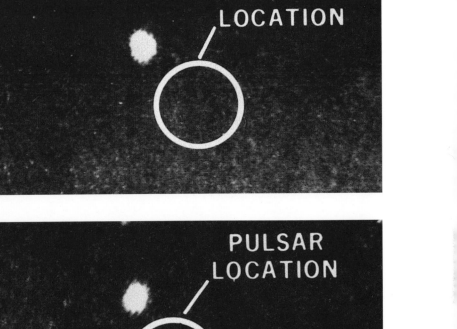

NASA

Radio astronomers discovered that the radio signal from the Crab nebula (right) blinks on and off 30 times a second. Visible when the radio signal is on and disappearing when it is off, this neutron star is about one third the mass of the Sun, but less than ten miles (16 kilometers) in diameter.

NASA

looking for evidence of black holes. In some ways, it would be particularly fitting if she were to find one. Black holes, if they exist, may well be the most powerful objects in the universe. Burbidge has already discovered the fastest-traveling object in the universe on record, a quasar. In fact, she has done it twice.

When she was first entering the field of astronomy, the British-born Burbidge says there was a feeling that few new discoveries lay ahead. "Back in 1940, there were some rash people who said that since

now we have a pretty good idea about the properties of stars and can look at the structure of galaxies, there's not much more to do," she reminisces. "But astronomy goes in bursts of discovery. Whenever you make dramatic improvements in a technology or get a new technology, you get these bursts of discovery." The new and improved technology that led to Burbidge's breakthrough was radio astronomy.

Like CCDs, radio astronomy was a serendipitous discovery of Bell Labs, which, in the 1930s, had an

engineer build a large antenna in northern New Jersey to track the origin of the static noise that was interfering with transoceanic radio telephone reception. The engineer, Karl Jansky, found that while some of the disturbance came from local and more distant thunderstorms, other disruptions came from a mysterious third source. After more that a year of close observations, Jansky concluded that he was receiving radio transmissions from a source near the center of the Milky Way. For the most part, as-

Quasar 3C273 (shown in an X-ray photograph, lower right), discovered about fifteen years ago, is a familiar sight to scientists.

tronomers ignored this epochal discovery, perhaps because it had been made by a nonastronomer.

During the Second World War, however, a British astronomer serving as a radar troubleshooter accidentally discovered radio transmissions from the Sun. Using the same antenna after the war, he discovered the first radio source outside the Milky Way, a galaxy that is a powerful source of radio energy—what astronomers now call a *radio galaxy*. Radio astronomy was well on its way to becoming an important new scientific tool.

By the late 1950s, radio telescopes could be used to pinpoint radio sources in the sky. And these sources turned out to be, when looked at with optical telescopes, radio galaxies. Or a least they did until in 1960, astronomers happened upon a radio source (named 3C 273 because it is the 273rd object in the third radio survey conducted by England's Cambridge University) that appeared as a faint bluish star. The radio astronomers dubbed it a quasistellar radio source, the optical astronomers a quasistellar object, but before long it was universally known as quasar.

At first, astronomers assumed that 3C 273 was a star in the Milky Way. But, in 1963, an examination of its spectrum forced a reevaluation of this hypothesis. With a spectrum more severely redshifted than anything that had previously been seen in the heavens, 3C 273 was apparently moving away from Earth at a speed of 31,068 miles (50,000 kilometers) per second. Based on Hubble's law, that the speed of celestial objects is directly related to their distance from Earth, this meant that the quasar was more than 2 billion light years distant. It also meant that if 3C 273 really was that distant, it had to be one hundred times brighter than the brightest radio galaxies known. Since then, some six hundred other quasars have been discovered.

Burbidge's distinction is to have twice participated in finding the fastest quasar. She discovered the first quasar in 1966, although that one only maintained its lead in the astronomical space race for a month. In 1973, she and a coworker sighted the still-reigning champ, OQ 172 (the 172rd object in the University of Ohio's quasar catalog), which is darting away from the Milky Way at faster than ninety percent of the speed of light, that is, at roughly 168,363 miles (270,000 kilometers) per second—an inconceivable hurry.

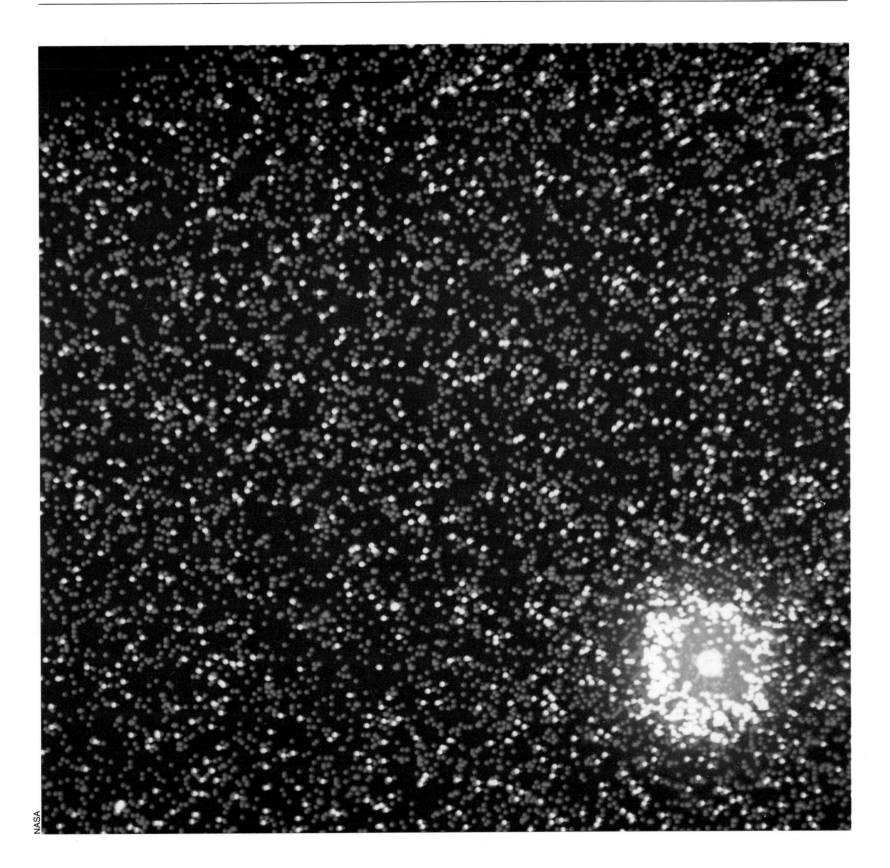

And what can we hope to learn about this from peering at quasars and distant galaxies? Nothing less than the infancy of the universe according to Hubble, says Bahcall of the Space Telescope Science Institute. "Simply by looking at galaxies with very high red shifts, galaxies that are very far away from us in the universe—looking at their structure, density, and distribution—we will learn a great deal about how galaxies were formed when the universe was young," she explains. "We can detect these galaxies from Earth, but we only see them as a speck. With the high resolution of the space telescope, we should be able to see their structures and measure their properties. We see them as they were very early on because it takes so many billion years for the light to get to us.

"It is as if we on Earth in order to understand the evolution of the human species could see in detail how life on Earth was some millions of years ago," she continues. "As if we could see a picture of the Earth a few million years ago, when there were dinosaurs; then a million years ago, when the first people emerged; then a hundred thousand years ago; then ten thousand years ago—a whole series of pictures like that. That is what the space telescope will be able to do. We will look at galaxies 5 billion light years away, 1 billion light years away, and so on, up to some millions of light years away. Comparing how galaxies were billions of years ago with how they are today should tell us a lot about the evolution of the universe."

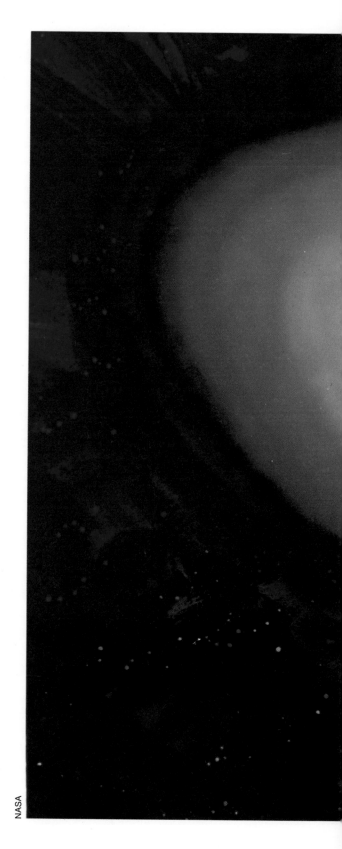

A disc-star located in the constellation Cygnus, shown here in an artist's conception, is believed by some astronomers to be forming planets.

NASA

IN CONSCIOUS EXPECTATION OF THE UNEXPECTED

In the United States, a young astronomy devotee can aspire to growing up to hold a title such as Leckrone's: space telescope instruments scientist. In the United Kingdom, by way of contrast, a monarchial tradition creates the possibility of grander sounding appellations, and any star-struck males in that country can dream of becoming an astronomer royal, a dream whose realization is made slightly easier by there being two astronomers royal, one for England and one for Scotland.

Astronomically minded women, however, will have to keep their ambitions in check. In 1972, when Burbidge, on leave from her teaching work at the University of California at San Diego, served as the first woman director of the Royal Greenwich Observatory, she was not given the honorific title of as-tronomer royal for England that traditionally goes with that job, a denial of the fact that women have done important work in the field of astronomy at least since the late eighteenth century.

At that time, the German-born Caroline Herschel followed her brother William to England to assist him first in his musical career—he was a church organist and composer; she was a talented singer—and then in his scientific endeavors. In 1781, William Herschel sighted Uranus, the first planet to be discovered since ancient times. The resulting fame secured for him an annual pension from King George III of £200 for himself and £50 for Caroline as his assistant. Although she for the most part limited herself to recording her brother's findings, she did in her spare time perform her own

NASA

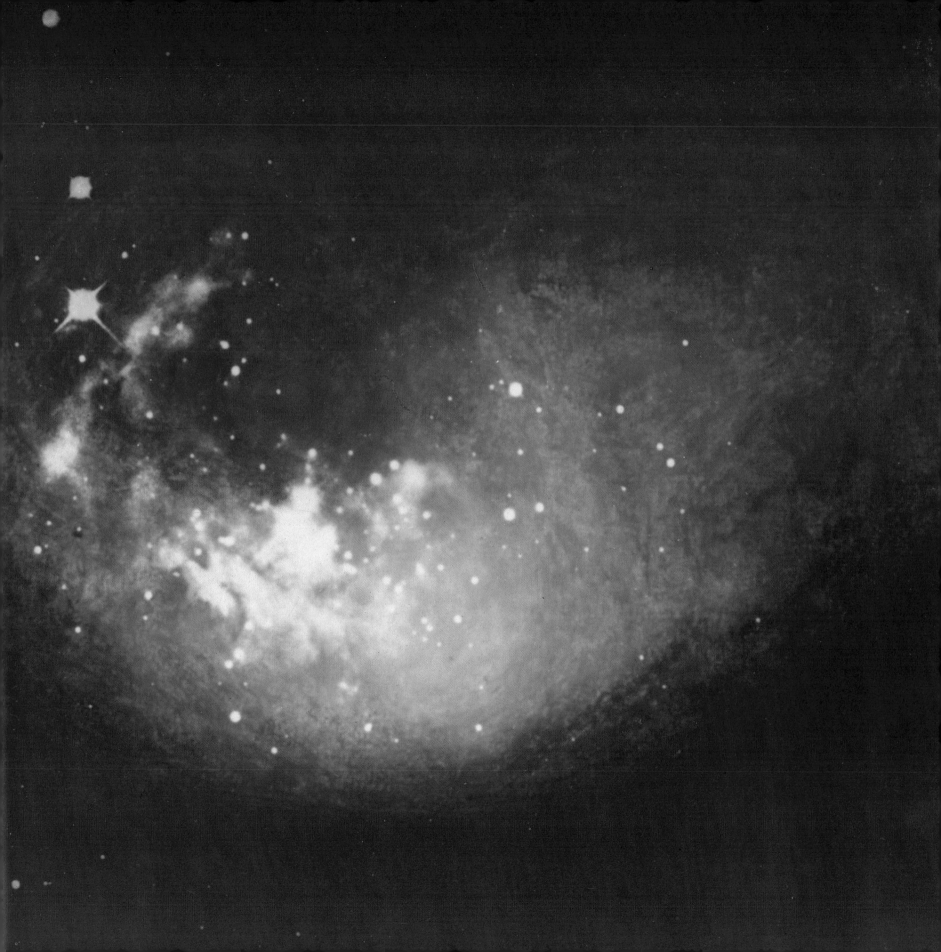

The greatest discoveries of the space telescope may well involve celestial objects as yet unknown to and unimagined by astronomers such as Orion Nebula M42 (previous page).

The Vela Supernova (right) exploded some 10,000 years ago, but we're still seeing the spread of its filaments from the blast. Its mysterious faint-blue rim encases the brighter red glow of its luminous insides. Slightly off center is Pulsar 0833–45, a bright spinning neutron star, and perhaps the remains of the exploded star itself. Malcolm Longair (inset), astronomer royal for Scotland and the only non-American member of the space telescope's working group, anticipates the time when the telescope will bring back even more spectacular data than the discovery of this supernova.

observations and is credited with discovering eight comets.

The title astronomer royal for Scotland accompanies the position of regius professor of astronomy at the University of Edinburgh and the directorship of the Royal Observatory, Edinburgh. It has been held since 1980 by Malcolm Longair. In his association with the space-telescope project, Longair also enjoys a post with a more bureaucratic, less aristocratic title. He is one of the four interdisciplinary scientists, the only non-American, participating in the space telescope's science working group, the main scientific advisory body to the project.

Longair has long been an active proponent of the space telescope. "My own view is that the telescope is really excellent value for the money," he remarks. "The spectacular thing is going to be the sheer quality of the data we get back. It will open up a new era in astronomy." He is particularly excited about the possibility that the telescope will change our entire view of the heavens. "You just can't have this sort of increased sensitivity and resolution and not obtain a radically new view of what the universe looks like." But, although Longair clearly agrees with Spitzer that the space telescope will be a great advantage for astronomy, the historical analogy he prefers to be made is not to Galileo but to the mid-sixteenth-century astronomer Tycho Brahe. Says Longair, "Tycho was one of the greatest astronomers of all time and in many ways the founder of modern astronomy."

Born to an aristocratic Danish family, Tycho was so impressed in his youth by the ability of astronomers to predict a partial eclipse of the Sun that he decided to devote himself to the field. Three years later, when he observed a conjunction of the planets Saturn and Jupiter that the astronomers had not predicted, he decided to devote himself to making precise and systematic observations of the night sky. His most spectacular single observation began on 11 November 1572 when he saw a bright star in the constellation Cassiopeia that had never been seen before. A few days later the star was shining as bright as the planet Venus, and then it began a slow fifteen-month fade to invisibility.

Through close observations, Tycho was able to prove the new star was not a comet and was as distant as the other "fixed stars," refuting the Aristotelian concept of a changeless universe and paving the way for the acceptance of the Copernican idea of a Sun-centered Solar System—an idea Tycho supported in a series of lectures at the University of Copenhagen in 1574. And based on Tycho's notes, contemporary astronomers know that the new star was what we call now a supernova—the result of an explosion in a star about ten times more massive than our Sun beginning to contract on its way toward star death. Known today as Tycho's supernova remnant, the remains of the star are faintly visible to the best optical telescopes and highly visible to X-ray and radio telescopes.

But Tycho's most important work was more meticulous, if less

© Royal Observatory, Edinburgh/W.J. Robertson (Inset)

© Royal Observatory, Edinburgh

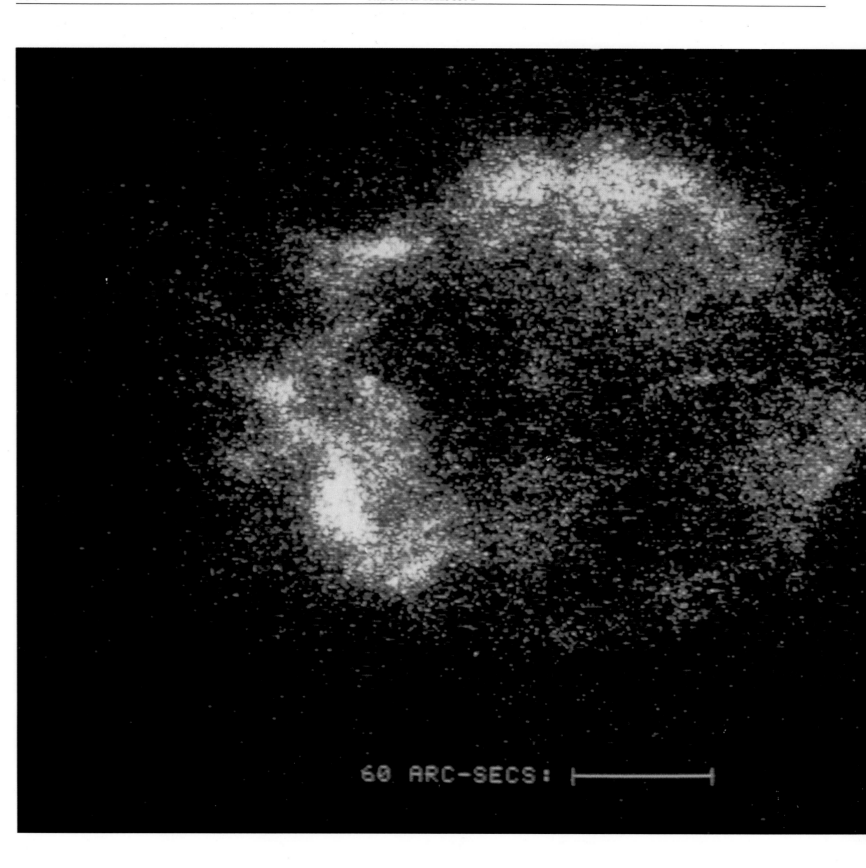

60 ARC-SECS:

Mary Evans Picture Library (inset)

NASA

spectacular. In 1576, King Frederick II of Denmark, fearing that his country's most famous scientist might settle in Switzerland, offered Tycho the funds to build an observatory and a 2,000-acre (810-hectare) island called Hveen on which to build it. The astronomer took his monarch up on the offer, and his "observations of the stars and planets, stretching over twenty years, achieved an accuracy ten times better than any previous work," comments Longair. "His data enabled [Johannes] Kepler to show that the orbit of Mars was not a circle but an ellipse with the Sun located at one focus. This conclusion could not have been drawn from the previous work because the data was too imprecise. In turn, Newton derived his law of gravity from Kepler's laws in one of the greatest intellectual achievements of all time. None of this would have been possible without Tycho's brilliant observations."

Will the space telescope in a similar manner provide the background for a revolution in twenty-first-century astronomy? Astronomers are reluctant to make extravagant predictions. They don't want to be remembered for their sightings of canals on uninhabited planets or imaginary star wobbles. But emblazoned on the side of a coffee mug Burbidge keeps on her office desk at the University of California at San Diego are words that express that attitudes of these scientists toward what they hope to find. The words, from the announcement Goddard issued back in the mid-seventies when it was looking for instrument proposals, read: "In conscious expectation of the unexpected."

An X-ray photograph reveals a supernova remnant in the constellation Cassiopeia. Brahe's (inset) observations of a supernova in 1604 led astronomers to abandon the Aristotelian idea that the stars were fixed and unchanging.

B I B L I O G R A P H Y

About the space telescope:

Bahcall, John N. and Spitzer, Lyman, Jr. "The Space Telescope." *Scientific American* (July 1982): 40–51.

Beatty, J. Kelly. "Space Telescope: Problems and Progress." *Sky & Telescope* (September 1983): 189–190.

_____. "HST: Astronomy's Greatest Gambit." *Sky & Telescope*. (May 1985): 409–414.

Godwin, Ira. "Space Telescope: 2 years late, $1.2 billion spent, and still counting." *Physics Today*. (November 1983): 47–49.

Hanle, Paul A. "Astronomers, Congress, and the Large Space Telescope." *Sky & Telescope*. (April 1985): 300–305.

Henbest, Nigel, ed. *Observing the Universe*. Oxford: Basil Blackwell and New Scientist, 1984.

Longair, Malcolm. (Interview with) "Malcolm Longair: Scotland's Astronomer Royal." *Sky & Telescope*. (June 1984): 516–518.

_____. "The Scientific Challenge of Space Telescope." *Sky & Telescope*. (April 1985): 306–311.

Lightman, Alan. "First Light: The Space Telescope," in Cornell, James and Carr, John, eds. *Infinite Vistas*. New York: Scribners, 1985.

Lubkin, Gloria B. "Perkin-Elmer ships 2.3-m optical space telescope assembly." *Physics Today*. (November 1984): 17–19.

Spitzer, Lyman, Jr. "Astronomical Research with the Large Space Telescope." *Science*. (July 10, 1968): 225–229.

Tucker, Wallace. "The Space Telescope Science Institute." *Sky & Telescope*. (April 1985): 295–299.

Tucker, Wallace and Karen. *The Cosmic Inquirers*. Cambridge, Massachusetts: Harvard University Press, 1986.

Waldrop, M. Mitchell. "Space Telescope in Trouble." *Science*. (April 8, 1983): 172–173.

_____. "Space Telescope (I): Implications for Astronomy." *Science*. (July 15, 1983): 249–251.

_____. "Space Telescope (II): A Science Institute." *Science*. (August 5, 1983): 534-536.

_____. "New Worries About Space Telescope." *Science*. (June 8, 1984): 1077–1078.

Watson, Gary. "Building the Space Telescope's Optical System." *Astronomy*. (January 1986): 15–22.

Background works on astronomy, astrophysics, etc.:

Barrow, John D. and Silk, Joseph. *The Left Hand of Creation*. New York: Basic Books, 1983.

Golden, Frederic. *Quasars, Pulsars and Black Holes*. New York: Scribners, 1976.

Harwit, Martin. *Cosmic Discovery*. New York: Basic Books, 1982.

Henbest, Nigel and Marten, Michael. *The New Astronomy*. Cambridge: Cambridge University Press, 1983.

McBride, Ken. "Looking for extrasolar planets." *Astronomy*. (October 1984): 6–22.

Trefil, James. *The Moment of Creation*. New York: Scribners, 1983.

Weinberger, Steven. *The First Three Minutes*. New York: Basic Books, 1977.

Whipple, Fred C. *Orbiting the Sun*. Cambridge, Massachusetts: Harvard University Press, 1981.

NASA

I N D E X